The Design and
Implementation
of Geographic
Information Systems

The Design and Implementation of Geographic Information Systems

John E. Harmon and Steven J. Anderson

WILEY

John Wiley & Sons, Inc.

For general information on our other products and services or for technical support, please contact our Customer Care Department within the United States at (800) 762-2974, outside the United States at (317) 572-3993 or fax (317) 572-4002.

Wiley also publishes its books in a variety of electronic formats. Some content that appears in print may not be available in electronic books. For more information about Wiley products, visit our web site at www.wiley.com.

Library of Congress Cataloging-in-Publicaton Data

Harmon, John E.
   The design and implementation of geographic information systems /
John E. Harmon, Steven J. Anderson.
      p.cm.
   Includes bibliographical references (p.)
   ISBN 0-471-20488-9 (cloth : alk. paper)
   1. Geographic information systems. I. Anderson, Steven J., 1964- ii.
Title.

G70.212.H36 2003
910'.285—dc21   2002032425

10 9 8 7 6

CONTENTS

ACKNOWLEDGMENTS

G IS is an extremely collaborative activity. Almost at the drop of a hat, GIS professionals will organize a conference, set up a Web-based list server, pull people together for discounts on training, establish a networking group, and generally do what they can to move the community forward. The standards for GIS certification in the United States, which are being worked out as we complete the manuscript for this book, explicitly include points for participating in and organizing events that benefit the entire community. Collaboration, making your data and documents open and accessible to other users, is what GIS professionals do, and we are proud to be a part of that community. But there is still real value in having a book, something between covers on the shelves of your office, that you can pull down to help you work through a problem. It is always interesting to go into the office of a GIS professional and, while making the customary chatter about backgrounds, look around on the bookshelves. What books did this person deem worthy of holding on to? Are there books besides software manuals? Which books appear to have been heavily used, pulled down and opened many times? A bookshelf says a lot about a person. So we have taken the opportunity to write a book that we hope will go up on a few shelves, but we could not have done it without the support of many in the community. Specifically, they are Gerry Daumiller, Natural Resource Information Center, State of Montana; David Dickman, SBC — Southern New England Telephone; Frank DeSendi, Bureau of Planning and Research, Pennsylvania Department of Transportation; Kevin Hanron, Charles River Technologies — Crimeinfo; Jay Heerman, Address Coordinator, Johnson County, Kansas; Neijel Huff, GIS Intern — Town of West Hartford, Connecticut; David Kingsbury, Intergraph Corporation; Barbara MacFarland, Metropolitan District, Hartford, Connecticut; Donny McElveen, Department of Transportation, South Carolina; Jeffrey Osleeb, Hunter College, CUNY; Martin Roche, Information Technology Director, South Carolina Department of Commerce; Jeffrey Roller, GIS Coordinator, Town of West Hartford, Connecticut.

Introduction

A geographic information system (GIS) is really nothing special. Like any information system, a GIS is an organized accumulation of data and procedures that help people make decisions about what to do with things. In a GIS these things have one characteristic that makes them at least a little special — their location is an important part of what they are. You as a human being are a special thing, but your basic self doesn't change a lot depending on where you are, although some might take issue with that. You may be one person at work and another at home with your family, but those are characteristics that differ depending on the role you have adopted, not where you are. The role of a parent brings out different characteristics that are not related to where you are being a parent but just to the fact that you are acting as a parent. But a parcel of land is completely tied up with where it is; you can't move it and have it be the same thing. A segment of pavement is where it is, and if you pick it up, move it and disconnect it from other segments of pavement, it becomes something different. People have been constructing GISs to manage and analyze types of things for which location matters for almost the last 40 years. In this book we discuss and explain the issues of design and implementation of systems that manage this type of information.

Who Should Read This Book

We have written this book to help practitioners design and implement multiuser, multiunit GISs to assist in their spatial decision-making. These are called corporate or enterprise GISs. We presume that you know what a GIS is, already have some experience with GIS, and have some ideas about what it can do. We also assume you are interested in improving the design of your existing GIS or building a well-designed system that will meet your needs. In our review of the existing material on GIS we felt that there was a missing piece in an accessible book. There is a lot of information on design and implementation in hundreds of needs assessments and database documentation files and probably thousands of public and private documents in the hands of organizations and consulting companies that have implemented GISs. But that information is very hard to come by and is usually specifically tailored for the particular application. At the other end of the generality scale, books on relational database design do not contain information on how to deal with spatial data, the "where" of things, and they usually are aimed at a business market with examples drawn from the world of commerce. Knowing how to design an inventory control or billing information system is a useful exercise, but it doesn't help a GIS practitioner in local government design a database to support planning and zoning activities. Our goal in this book is to deal explicitly with the issues of spatial data in designing and implementing a GIS and to provide examples useful to the GIS community.

People become involved in GIS through all sorts of complex pathways. Some come to it through a background in civil engineering and computer assisted drafting (CAD), whereas others may develop an interest through remote sensing of the environment. Many come to GIS from backgrounds that are not technical or mapping related; they just recognize that a GIS might help them do their jobs better. Others come from backgrounds in planning, geography, public health, surveying, property assessing, public safety, indeed from any of the dozens of application areas that can benefit from GIS. Because there are so many areas of human activity that can benefit from GIS, we have taken great care to keep the discussion general enough to be helpful without focusing on any particular application area. The *How They Did It* sidebars are our attempt to bring in some specific concerns around data, applications, and software that organizations need to consider. By no means do they exhaust the application areas, the data issues, or the software requirements for organization, but they provide some experiential background for the design and implementation process.

What Is a Geographic Information System?

Having said that we assumed you knew what a GIS was before picking up this book, we feel an obligation to define it anyway. So, to repeat what has been circulating in the information technology (IT) and GIS literature, vendor brochures, countless figures, and PowerPoint slides, a GIS is composed of

- *People* — the users of the system
- *Applications* — the processes and programs they use to do their work
- *Data* — the information needed to support those applications
- *Software* — the core GIS software
- *Hardware* — the physical components on which the system runs

Versions of these components of a GIS have been circulating in the literature and vendor brochures for years. They certainly predate GIS specifically and are applicable to any information system. Although it is common to represent the five elements, or components, in some kind of mandala diagram where everything appears to be connected to everything else, we do not find that very useful, so we have added the key relationships. Figure 1.1 should be read as a sentence, and it proceeds from the most important elements to the least important elements.

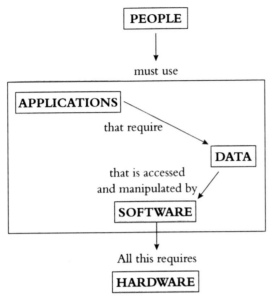

Figure 1.1 Components of an enterprise GIS.

The people are the most important component, although some would argue for the data. Information systems, geographic or not, spring from the needs of people in organizations to do work, answer questions, and generally interact with the world and the people and organizations in it. An information system is supposed to support the work, to make it quicker to do with more consistent results, and to provide high levels of confidence in the output. The process of design and implementation of a GIS begins with people and their needs and ends up with applications in the hands of people who do the work. The entire system exists to support them and their tasks.

The applications come next in the hierarchy because they define the work that needs to be done. In organizations people need to create all kinds of reports, make all sorts of decisions, and generally apply their skills so the work gets done. The processes they develop to do these things are the applications. Some applications are routine and get done multiple times a day, whereas others are less routine but get done with some regularity, and then there are specific analytical applications that might have to be accomplished only rarely or even just once. The applications arise out of the mission and goals of the organization. In any information system you need to know what applications the system will be expected to support.

Applications require data to work. You can't generate a map of sales potential or customer locations without the appropriate data tables necessary to create that type of output. These tables will reside in a database (possibly more than one), and the system will require software to access, manage, and manipulate the data so that the application can generate a useful product. The data support the application, and if there were no software for data storage and retrieval, the application would have to get done somehow. Before the 1970s and the advent of relational database management systems (RDBMSs) in business and government, bills got generated and paid, and locational decisions were made. The process was a little more cumbersome, but it worked. So an application can work without software so long as the data are arranged in a useful way. That is why data are more important than software. The triad in the center of Figure 1.1 of applications/data/software represents the core of the information system. Ideally, it should work regardless of who the people are (if the application is well designed) and be flexible enough to work on whatever innovations in hardware come along, even in the absence of any hardware. That is why hardware is at the bottom, the least important element of the information system.

Corporate or Enterprise Geographic Information Systems

The type of GIS we discuss in this book is a corporate, often called enterprise, GIS (Harder 1999; Von Meyer and Oppman 1999). An enterprise GIS is one that is designed to meet the needs of multiple users across multiple units in an organization. Although many organizations have a GIS in one or more units, they have been built to support the needs of those units only and may be of little utility to other departments in the organization. It is common to find an organization where a marketing department has one desktop GIS license to support its work, the engineering department has a GIS with CAD at its core, and perhaps other scattered, lightly or heavily used systems are in different units. There is much duplication of data and little congruence of the data sets, duplication of applications, variety in standards for the output of the systems, and generally a unit-centric view of GIS and what it can do.

An enterprise GIS is built around an integrated database that supports the functions of all units that need spatial processing or even mapping. That database, whether centralized for real-time access by all users or replicated across many computers, is the engine of the enterprise GIS. In a well-designed system, users in the departments where GIS already existed will interact with the GIS in ways that are not much different from what they had been doing. New users will interact with the system with custom-designed applications that use the centralized data. The system will no longer be a particular department's GIS but will be the organization's GIS.

This kind of corporate, or enterprise, GIS is different from a single-unit or project-oriented GIS in several ways:

- ◆ *Data are standardized and redundancy is reduced.* In municipal government, a type of organization that is an excellent candidate for an enterprise GIS, an assessment of information needs will almost always reveal that many different units of the government maintain information on addresses. But there will be no standardization of address composition; that is, some units may store addresses in a single field of a data table, others break it into street numbers and streets, and a still others might use the address parsing scheme that came with a packaged information system. As a result, there is a lot of duplication and confusion about what an address is and how to store it. An enterprise GIS for a local government may have standards and a nonredundant master data table of addresses.

- *Database integrity is maximized.* When people start using information and modifying data in databases, what was once clean and accurate data has a way of getting dirty and inaccurate. Names are misspelled, addresses incorrectly recorded, records deleted that should not have been; the list of things that can happen to corrupt data is very long. An enterprise GIS will have safeguards and procedures to minimize that kind of data loss. All organizations run on a supply of accurate and timely information to make decisions that will move the organization forward, and it should be as good as possible. The progression of data in information systems, GIS included, has been from databases on mainframe computers maintained by centralized staff to distributed databases on individual desktop computers, and now, full circle, back to central data servers. The data are too important and need careful watching and maintenance.

- *Units come together through the database.* In a complex organization there are many different departments or units with unique missions and goals. When they come together in the creation of an enterprise GIS, they become aware of other units' needs for information and begin to see their own needs in a different light. People in almost all of the organizations that undertake and successfully complete the implementation of an enterprise GIS will speak positively about the benefits to the organizations that go beyond the cost and times savings the GIS might provide. An enterprise GIS database is a gathering point for different units, and how they come together in its creation and maintenance usually benefits the organization in some unexpected ways.

- *There is a consistent look and feel to output.* In the process of design and implementation of this kind of database it is necessary to set standards and requirements on what the output from the database will look like. This results in a consistent-appearing output that is usually important to top management. Typically, management does not want any individual units of the organization looking too much out of step with the others, and an enterprise database with standardized outputs makes uniformity much more likely.

- *Geographic information costs are centralized.* A common problem that managers encounter when they are building a case for an enterprise GIS within the organization is that the current costs of obtaining and using geographic information are diffused and hidden within the budgets and operations of many different units. Anyone who tries to conduct a cost-benefit analysis around a GIS discovers this. It is comparatively easy to calculate the costs of implementation but much more difficult to compare those costs to ones presently being incurred. By taking an enterprise GIS approach, you make the conscious decision that the costs of geographic information should be borne by the entire organization, not the individual units. Cost centralization with benefit dispersion allows the organization to use its resources more effectively

It may sound as though every organization in the world that deals with geographic information would be foolish not to rush right out and develop an enterprise GIS, but that is not the case. The software vendors and GIS consultants would probably like all organizations to feel that way, but they won't. It is largely a matter of size and complexity. Very small water companies that have a small number of wells and provide water service only a few hundred homes will probably get by with a single license of a CAD system and a set of spreadsheets and a desktop database management program. To expect a small company with a staff of five to invest in multiple software licenses, multiuser database management systems, and all the support those things require is unrealistic. A water utility serving several hundred thousand accounts across multiple political jurisdictions, though, is a prime candidate for this kind of GIS implementation.

Size and complexity are not necessarily the same thing; some small organizations can be quite complex. This is the case with most municipal governments in which there tend to be large numbers of different departments or units, sometimes lightly staffed, that have distinctly different missions. This kind of organization, even though it is not particularly large, may benefit greatly from a well-designed enterprise GIS.

The process of design and implementation of an enterprise GIS is complex and can take up to a year or more depending on the size of the organization and the amount of geographic data which need to be incorporated. But unlike some complex planning processes, this one is relatively straightforward, without many feedback loops and decision points. The process consists of a set of one-time steps that you need to get through and some continuing steps (see Table 1.1).

One way of thinking of the design and implementation process is to identify the products that come out of each stage. The needs assessment/requirements analysis is discussed in chapter 2, but it should always precede any attempts to prepare a strategic plan. The needs assessment will give you an idea of where you are; the strategic plan is an outline to get to where you want to be. Our discussion of the second plan, the implementation plan, sometimes comes along with the needs assessment/requirements analysis if the organization is already sure it wants to jump into GIS. We discuss that plan in chapters 7 and 8. The third plan, maintenance, is discussed under system management (chapter 10, Managing the System the Maintenance Plan). Those four documents — the three plans and the needs assessment/requirements analysis — are basic for the implementation of a system that could take several years and hundreds of thousands of dollars if not more. Additional documents might include cost/benefit analyses, requests for proposals (RFPs) for consulting work, training manuals, data dictionaries, and so on. A complex GIS generates a lot of paper and that is a good thing. It means other people, particularly new staff, can quickly come to understand the system, how it was developed and how it is structured. A GIS where the documentation resides in the heads of one or two key staff people is on very thin ice.

Table 1.1 Steps in the Design and Implementation Process

Steps	Central Questions	Primary Outcomes/ Products	Who Is Involved? (Staff or Consultants)	Secondary Goals	Where Discussed in the Book?
		One-Time Steps			
Needs assessment/ requirements analysis	What are the current and future needs for geographic information? How do people currently use geographic information and how would they like to?	Needs assessment report. This is a central document to guide the implementation plan phase.	Potential users and managers of using departments and departments that may later become users.	To build support at front-line and mid-management levels.	Chapter 2 – Needs Assessment/ Requirements Anaylsis
Strategic plan	How will GIS further the mission of the organization? How will an enterprise GIS fit within the organization?	Strategic plan, with rough time-table, agreed upon by all stake-holders, not just top management.	Top and mid-level management plus core GIS committee.	To build support in top management.	Chapter 1 – The GIS Strategic Plan
Implementation plan	What are the phased steps we need to take to implement an enterprise GIS that will meet our needs and further our goals?	Implementation plan with detailed timetable, including a decision on high-level design questions.	Full GIS implementation committee with technical support.		Chapter 2 – Pulling the Needs Together, Chapter 7
Design phase	What data tables, fields, and initial applications are required to meet user needs and how should they be arranged?	Database schema, data dictionary, applications, flow charts.	Technical staff.		Chapters 3, 4, 5
Implementation phase	What processes will we use to populate the schema and implement the initial applications?	GIS database and selected applications.	Technical staff and front-line users.		Chapters 6, 7, 8
Pilot project	Where and when can we test this system?	Output information for pilot project.	Selected front-line staff and management.	To quickly show everyone, particularly management, that it will work	Chapter 7 – Perform a Pilot Project

continued

Table 1.1 *(continued)*

Steps	Central Questions	Primary Outcomes/ Products	Who Is Involved? (Staff or Consultants)	Secondary Goals	Where Discussed in the Book?
Continuing Steps					
Application development	How can we add to the existing set of applications and improve existing ones?	Updated procedure manuals.	Technical staff working with front-line users.	Convincing reluctant users of the system's value.	Chapter 9
Maintenance and upgrade plan	How will we keep the data, software and hardware current?		Technical staff with front-line management.	Ensuring ongoing support.	Chapter 2 – Maintenance, Chapter 10 – Managing the System – The Maintenance Plan
Training	Who needs what kind of training in what applications?		Technical staff and skilled users.	Convincing reluctant users of the system's value.	Chapter 2 – Training
Evaluation	After a reasonable training and working period, is it working as advertised?		All users.	Convincing top management of the wisdom of their decision.	Chapter 10 – Evaluation

The GIS Strategic Plan

Strategic planning is an activity with which most complex, multiunit organizations are familiar. Over time, the organization may have prepared many strategic plans, overall plans around the goals, and specific plans for projects and other important changes the organization has gone through. A strategic plan for GIS is therefore just another working out of a familiar process and should not cause a lot of anxiety, although midlevel management and front-line staff often feel a little put out about participating in yet another planning exercise. A strategic plan for GIS is a document of interest principally to top-level management and outside organizations. Top-level managers needs a document they can refer to when difficult decisions need to be made; is this something that fits our strategic plan or not? In some ways an important function a strategic plan plays is as a backstop for decisions that management may have to make. The reason for a positive or negative decision in some area then becomes not "because I said so," but "because this proposal does not mesh well with the strategic plan." The strategic plan for an enterprise GIS should contain at least the following:

- *A concrete discussion of how an enterprise GIS fits within the existing mission statement of the organization.* Front-line workers and midlevel management in any organization cringe when they are asked to contribute to discussions around the mission statement because they feel, sometimes correctly, that it bears little relationship to what they do on a daily basis. So the people participating in this GIS strategic planning process need to be reassured that the purpose is not to rewrite the mission statement but to demonstrate how a GIS would support the existing statement of mission. Top managers in most organizations really believe these are important documents; to many they represent the condensation to a small number of statements what the organization does and why it does those things.

- *A tentative and light discussion of how GIS is going to fit inside the organization with a recognition that the design and implementation process may require modification of the plan.* A GIS strategic planning committee may decide that creating an entirely separate GIS unit parallel to but separate from the existing information technology department is a good idea. In the course of design and implementation, however, staff may pick up skills, money may get short, leadership may change, and the recommendation may change to something else. That could always happen. But all involved staff deserve to have an idea of where it is going to fit within the organization.

- *A timetable with checkpoints.* A plan without a timetable is a dangerous thing because it gives management no way to monitor progress. The implementation process can take more than a year, so there need to be checkpoints along the way to monitor progress.

How They Did It – State of South Carolina Strategic Plan for Statewide GIS Technology Coordination

The strategic plan for GIS coordination in the State of South Carolina is a good, if large and complex, example. From the resources available on the World Wide Web, agendas, minutes, and the documents produced, you can reconstruct at least the official process of this committee. The committee was established by an executive order of the governor. David Cowan, a university professor, had prepared some principles for statewide GIS coordination in January 1996, and the committee had its first meeting in July 1997. By January 1998 the committee, which was composed of more than 30 people at its height, was beginning the needs assessment process. A consultant was hired and the second draft of the needs assessment, a 224-page document, was completed by the consultant in July 2000. The needs assessment looked at state, regional, and local governments; universities; and other stakeholders; in all, more than a hundred different organizations participated in some way in the needs assessment. The linkages between the different stakeholders, their missions, and how geographic information was important to meeting that mission were clearly presented in the needs assessment.

Along the way in this process the committee also had to deal with other issues that just arose, and because they were in place, they become their problems. One was the relationship between GIS and the surveying professions and how that would work out in the South Carolina legislature where the committee had to work quickly to ensure that professionals other than surveyors would be able to work with parcel layers in a GIS. There was also relationship between the committee and an existing Statewide Mapping Advisory committee that had been in place for more than 20 years. The principle focus, however, was on the needs assessment and strategic plan.

Throughout the 3-year process the committee went through several different structures, tried to hire a coordinator, finally found one in a member of the committee, but kept the same consultant during the process. Funding was from the good will of 10 state agencies, each of which could find $10,000 for each of 2 years for a $200,000 budget. Virtually all of the money was spent on consulting contracts and expended over an 18-month period. The committee consulted with senior GIS staff from all state agencies with GIS capabilities. Martin Roche, who coordinated the committee, estimated it took 10 to15 percent of his time over the period, 5 to10 percent of the time of four other members of a steering committee, and less than 5 percent for the other members of the committee. At around the time of the completion of the draft, the sponsoring governor lost his election, several key supporters retired, and the new governor appointed a new committee. The committee, though it officially still exists, has been dormant since the completion of the plan. This complete draft was completed in January 2001.

The strategic plan consists of two separate documents, one of 24 pages (GIS Background and Business Case) and a second of 60 pages

(Strategic Foundation and Work Program). The things that happened along the way — loss of key political support, loss of key committee members, getting sidetracked on important but ancillary issues — are things that can happen in any lengthy process and clearly had an effect on South Carolina's plans. This is probably as complex and lengthy a process that one could imagine, but it contains the outline of what any reasonably complex organization needs to do as it tries to coordinate GIS activities. Currently (May 2002), the plan is "just sitting there due to budget cuts. It is an ambitious plan and we are hoping for implementation in the next budget year." (Martin Roche, personal communication)

- *If the implementation is to be phased (i.e., brought into certain units first and later diffused to the entire organization), those units need to be identified in the strategic plan.* The pilot project will come from one of these early units, and the data layers and applications they need will be the first to be developed, so there needs to be organizationwide understanding of what the phasing pattern will be. This may lead to difficult decisions if there are managers from many units who want to be first, but it can also lead to easy decisions if there are managers from some units who are skeptical and would prefer to come later.

- *Initial decisions on what existing staff will do in the design and implementation process and what functions the organization will hire consultants to do.* A few organizations are able to do the entire process with existing staff, but these are usually organizations with considerable technical mapping and database experiences (e.g., large utilities). Most will find that a mix of in-house staff work and consultant contracts will provide the best mix of skills and get the GIS implemented more quickly without taking away too much time from the work that already needed to be done. The strategic plan is not the place to detail the RFPs for the consultant portions or to choose the consultants, but it is the place for management to come to agreement over what the mix will be. As part of this, if a GIS implementation committee is going to be a part of the process, the strategic plan is the place to lay out its makeup and responsibilities.

- *Some statement of the resources the organization is willing to contribute to the process.* This should not be a detailed budget at this point but some recognition of the costs and how the organization plans to pay for them. For large organization such as utilities that cover large territories the time to implement a GIS may be measured in years and the costs in millions of dollars. Ballpark estimates of the time and financial resources needed should be part of a strategic plan.

That is about it. It sounds simple but can be difficult. Finally, South Carolina notwithstanding, the plan should not be overly long, and there should be a one-page summary of the entire document as well. Plan for sufficient time to prepare the strategic plan; it is a critical document for building high-level management support for GIS implementation.

ADDITIONAL READING

Antenucci, J. C., K. Brown, P. L Croswell, M. J. Kevany with H. Archer. 1991, *Geographic Information Systems: A Guide to the Technology.* Kluwer Academic Publishers: Dordrecht, Netherlands.

Aronoff, Stanley. 1991. *Geographic Information Systems: A Management Perspective.* WDL Publications: Ottowa, Canada.

Harder, C. 1999. *Enterprise GIS for Energy Companies.* Environmental Systems Research Institute: Redlands, CA.

Huxhold, W. E., and A.G. Levinsohn. 1995. *Managing Geographic Information System Projects.* Oxford University Press: New York.

Huxhold, W.E. 1991. *An Introduction to Urban Geographic Information Systems.* Oxford University Press: New York.

Information Resource Council. 1997–2001. "Geographic Information Systems." Columbia, SC: Office of Information Resources. state.sc.us/irc/committees/gis/gis.htm.

Korte, G. E. 2001. *The GIS Book,* 5th ed. Onword Press: New York.

Longley, P. A, M. F. Goodchild, and D. W. Rhind. 2001. *Geographic Information Systems and Science.* John Wiley & Sons: New York.

Obermeyer, N. J., and J. K. Pinto. 1994. *Managing Geographic Information Systems.* The Guilford Press: New York.

Von Meyer, N. R., and R. S. Oppman. 1999. *Enterprise GIS.* Urban and Regional Information Systems Association: Park Ridge, IL.

INTERNET RESOURCES

GIS frequently asked questions:
census.gov/geo/www/faq-index.html

Some GIS portal or gateway sites:
ciesin.org/gisfaq/gis_gateway.html

directory.google.com/Top/Science/Social_Sciences/Geography/Geographic_Information_Systems/

Before Design: Needs Assessment and Requirements Analysis

Before you begin the implementation of a GIS, it is important to establish an essential groundwork that will make success with the implementation much easier to attain. The following factors need to be planned for in this crucial planning stage:

- Making sure that the proper individuals in your organization are involved
- Educating users
- Gaining commitment from management for the implementation
- Setting realistic expectations of users and management
- Understanding the needs and requirements of the user base
- Evaluating existing geographic data that is in use
- Assessing costs and benefits of the implementation
- Developing a strategic plan for implementation

Each of these considerations is discussed in the following sections.

Organizational Involvement

Having the proper individuals in the organization involved in the process of planning an implementation is a critical first step. One of the most common mistakes in planning the implementation of a GIS is to involve just management-level individuals in the process. Although managers are very important in the process, it is also equally important to have the staff who work for the managers involved for a number of reasons. First, staff-level individuals usually have a better understanding of the issues that are dealt with on a daily basis and the accuracies or inaccuracies that exist in the data they use. Second, staff will likely be more frequent users of the system than a manager would be because they typically deal with more of the day-to-day issues that can be solved using GIS than a manager does. Third, because staff will likely be more frequent users of the system, it is important to start building a feeling of involvement in the decision process early on in the implementation so that buy-in and acceptance can be achieved as early as possible in the process. Many new technology users often feel that if they are not part of the planning process, what is being developed will not work or will not be useful to them. If they are involved in the process, they can provide input to its design, see their feedback responded to, and take pride in ownership of the system as it evolves.

An effective way to involve users while still keeping control of the implementation path is to create two groups of users, or committees, that are involved in the decision process at varying levels. One group would consist of administrators or managers (often referred to as an oversight committee) who would be responsible for making policy decisions, setting security rights, making budgetary decisions, and rendering the ultimate decisions on the implementation path. The second group would include all of the same people in the first group and more of the end users of the system. This group would discuss operational issues of the system, discuss problems or issues with the data that are being developed or used, and recommend enhancements that could improve the system's functionality.

This group will also be very effective at educating other users and increasing the group's capabilities as a whole. Often one user will be experiencing a problem, and another user will know how to solve the problem. At each meeting of this group there should also be a designated presenter. A presenter can be a staff member from within the agency who has just completed a project using the system and can demonstrate the use of the system. A presenter can also be an invited guest such as a consultant, a staff member from a neighboring agency, a GIS coordinator from a utility company that serves the geographic area, or a software vendor. Much can be learned from what others are doing.

One of the other benefits that comes from involving users in this type of a committee is that it builds support for the system. As users spend more time in the planning stages of an implementation, they take much more ownership for the system. Ownership translates to support for the system. Support at all levels of the system is very important to its success. Invite management to these meeting so they can hear the users talking about what they are going to use the system for or how they have used it to solve problems. Let top managers hear reports from users who have seen implementations in other agencies, and they will realize that their company is not the only one that is doing something like this.

Another type of presentation can be a recap from a conference. Users who are sent to an off-site user conference or association meeting can be instructed to take notes and then make a presentation to the user group upon their return. This allows more users to benefit from the event. It also helps assure that the user who goes to the conference attends sessions that can be beneficial to the group as well.

In today's society human nature is to want what others have. There is a tremendous amount of competition in the private sector that has been carried over into the public sector. Competition for economic development purposes is often a driving force behind what makes agencies implement a GIS. With good data about itself, an agency can present itself as the place businesses want to come to. There have been many stories of where a GIS has been used to promote an agency, and its use has been the deciding factor when a business decides to come to that location. Learn about these success stories and make sure management is aware of them. On the other hand, GIS has also been used quite broadly for predicting the effects of what would happen if an agency was built out based on its present zoning. Build-out analyses have become common ways for communities to look at their future and make changes to regulations and constraints to assure that the community grows in the ways it wants to. Many agencies do not want to lose their present character, and GIS is an excellent way to visualize and understand what will happen if rules and regulations are not changed.

Need for Education, Support, and Commitment of Management – Corporate Implementation Takes Time

Planning a system as complicated as a GIS takes more than just sitting down in a room and outlining the steps that need to be taken, the data that are needed, and the software and hardware that need to be purchased. The process of planning a system starts with the need for a complete education about the capabilities of systems based on present-day and near-term technological advances. There is a need to know what works and what doesn't, pitfalls that have happened in previous implementations and how to avoid them, and what can be expected for time frames and results along the way. This education process can take many years to acquire, but there are a number of ways that this learning curve can be accelerated.

The first way is to visit agencies similar in size, or slightly larger, that have already implemented a system and see it in action. Ask the users what they like, what they don't like, what problems they encountered during the implementation, and what they would do differently if they could do it all over again. Don't just visit one place; visit a few so that you can get a better understanding of what the more common or consistent problems have been. Where the consistent problems don't exist, find out why. What did that agency do differently that allowed it to circumvent the problems others had? Take notes, especially when you see unique components that you like or dislike. In addition, encourage other potential users, especially those who are skeptical about the implementation, to meet with other agencies that are using the technology in the same roles they perform. Learn from other's mistakes.

Another excellent venue for learning about the technology is user conferences. Go to conferences that users who have already implemented a system attend to learn more about what can be done. Many events have presentations by end users or consultants who provide solutions in specific areas in which you may be interested. Talk to attendees of previous years' events to see which conferences are well attended and get good reviews; attend those events. Most conferences are organized around tracks that are geared to attendees who are just starting a system, have intermediate systems, and have advanced systems. Because you are just starting, attend mostly basic presentations, but don't be afraid to attend an upper-level session if the topic seems to fit you very well. Remember, you are planning for your future, not just what you can do today.

Another excellent source for educating yourself and other users is to invite consulting companies to present their approach to planning and implementing systems. Experienced consultants usually have very good presentations put together for these types of educational talks and often have multiple systems that they have developed for other agencies available for demonstration purposes. Another option is to visit the consulting companies' offices to see the different projects they are working on during the development process. Ask them to explain what and how they are doing on the project on which they are working. Again, learn from others to avoid mistakes.

Manage Users' Expectations – No Unrealistic Promises

Another important issue to deal with early on in the planning stage of an implementation is to manage the expectations of the users. Implementing a system takes time. Most well-planned, well-executed implementations take 2 to 3 years before they reach a mature level. What is important in the design stage is to design into the plan a series of milestones along the way so that success can be achieved and demonstrated. As departments are brought on line, it is important to provide small achievements along the way that will continue to build support. A good rule of thumb is to have at least one new function added to the system every other month. Many implementations fail because users' expectations are set so high at the onset of the project that when the goals are not met, they lose interest. If months and even years can go by without showing any new progress, support dwindles. Support takes a long time to build but is very easy to lose when promises are made that are unrealistic and cannot be met.

Needs Assessment/Requirements Analysis

A GIS is a very complex system that is composed of business processes, data, people, software, and hardware. The most critical stage of implementing a system is the planning stage. The first step in this stage is what is commonly referred to as a needs assessment. As the name infers, a needs assessment is a process in which you determine what all of the needs are for the users of the system. This includes determining the users who will benefit from the technology and which functions they perform will benefit the most, the applications of the technology in these functions, the data that will be needed to support these applications, the most appropriate software to support these applications using this data, and the hardware that will be required to support the system.

Performing a needs assessment can be a very time-consuming process and can result in erroneous results if not done in a systematic way. To understand how to plan for a system one must understand how systems have evolved in recent years. In the past, GISs were often started in a single department or division of an agency, and as these implementations progressed, a member of the user group acquired general knowledge of the technology, and this person became the champion of the system. The champion developed what was necessary to support the needs of his or her department, and the resulting systems were often very focused around this department's processes. Many systems failed because the champion of the system left the agency or was too busy with other duties to properly maintain the system, and it languished. In addition, the total needs and benefits of other departments in the agency were often not met because of the system's departmental focus.

In some cases, after success was had in this department, other departments would join in on the implementation, and it's use would grow department by department. Because this was more of a random than a systematic approach, it would work, but the implementation was not often the most cost-effective approach, and much time and money was often wasted. The following sections describe how one can plan to implement a system. There are many ways to approach this process, and this is just one way that success has been achieved.

Assessing the Current Users

The single most important component of the system that must be assessed is the users of the system. Because of the breadth of the processes that can be automated with this technology, there are many types of users who may touch or use the system over time. The most critical component of determining where to start is to determine which users will get the most value from the system. There are two primary classifications of users who can benefit from GIS technology. The first is users who are already using some sort of mapping in their day-to-day functions. These users are usually fairly easy to find because they are surrounded by some sort of hard-copy maps already. Assessors have tax maps, engineers have design plans, planners have zoning and land use maps, conservationists have wetlands and soils maps, and public safety staff have district, beat, or zone maps. Each of these users can benefit from the technology by automating these maps and making them more consistent and accessible.

The second classification of users can be much harder to identify. These are those who can use the technology to solve a problem, but the final output may not be a map. An example of this is routing inspections for an inspector. The technology can determine how to most efficiently get from one inspection to the next, and then the next, and so on, but the result may be printed directions, not an actual map.

To identify these users you must review their processes and determine whether there is a spatial component to what they do and then determine if GIS can be used to automate the process. Another example of such an application is document retrieval. Many departments have index cards or paper sketches of features that they retrieve based on an account number or some other type of identifier. These documents can be scanned and organized by linking them to a geographic feature with which they are associated, and GIS technology can then be used to retrieve the scanned version of the document once the associated feature is selected.

An efficient way to determine both of these needs is to first educate the users about all of the different basic functions that a GIS can perform. Once they are educated, the functions they perform and the amount of time they spend performing each function can be documented. These functions can then be analyzed and classified into different types of potential applications, whether or not mapping is used in the function and whether or not there is a spatial component to the decisions they are making.

Categorizing Users

Once the departments that will benefit from the technology are established, the next step in the process is to assess the people who use the system. The most efficient way to analyze the potential users and plan for their needs is to classify them into a series of groupings that are determined by the type of functions they perform in their regular work duties. Each user category must be well defined so that the applications, supporting data, and user interfaces that will be designed will support the functions that they perform. Care must be taken to develop groupings that will result in applications that provide all the capabilities necessary to support these roles without providing overfunctionality that will be confusing and not needed by other categories of users. The following sections describe typical user categories and the typical functions they perform. This section on users relates directly to what people are doing overall with geographic information right now. Later, in the early management phases of the GIS, you can assign particular roles that relate to how they will use the new database.

Director, Head, or Manager

A director, head, or manager is typically an end-user of the products and information created by other users of the system. Reporting and visualization of the data in the system is most important to this classification of a user. He or she is interested in knowing how many tasks have been completed, how much money has been taken in, or what the status of an activity is at any point in time. In many cases this person is not an actual user of the system; he or she asks questions of a subordinate, and the answers are provided.

The most efficient way to determine what functionality is required for this user is to review existing periodic reports such as weekly, monthly, or quarterly reports that this individual uses and determine whether the system can reproduce the same results.

Professional-Level or Technical Staff

Those at a professional or technical level typically perform technical reviews of activities based on their specialized expertise. A few examples of these types of users are engineers, planners, and scientists. These users usually need full access to all of the details of an activity and are typically responsible for performing technical reviews of activities to ascertain whether or not the activity is in conformance with the appropriate rules or regulations. In larger organizations these users do not typically input the details of an activity; they use the information provided by a subordinate. They are responsible for assuring that the appropriate reviews have been performed and the status of the activity is current. The details of activities such as date information, testing results, and measurements are usually very critical to this user. In addition they need to be able to query historic data, generate summary reports, and perform comparative analysis for use in defense of decisions they make.

The most efficient way to determine what functionality is required for these users is to review existing reports and checklists used and generated by them and determine whether the system can track the information used and generate a report containing the same information upon completion.

Administrative Staff

Administrative staffers are some of the most important users of a system. These users are the ones that are typically going to spend more time inputting and manipulating the data in the system and generating the reports needed for the staff they support. They need full access to perform data entry and data manipulation and to generate reports and maps.

The most efficient way to determine what functionality is required for these users is to review existing letters, mailing labels, file folders, maps, and reports generated by them in the current process and determine whether the system can assist them in the process. In addition it is important to review whether or not the information that is input by these users can later help the processes performed by any of the previously discussed users. Automating these processes can actually add time to the short-term process, but the benefits would then be seen in the overall process.

External Professionals (Engineers, Architects, Attorneys, Developers)

External professionals are a group of end-users who by nature will become expert end-users of a system like this. They will become experts because of the frequency at which they use the system. They are typically highly educated and knowledge-able of the process, able to interact with local government officials, and want to perform this function as efficiently as possible; therefore, they will to gravitate toward access to and use of systems like this.

They will want the ability to input data into the system themselves rather than fill out paper forms. They will require searching capabilities to review past activi-ties and reporting capabilities to print out what they have found and the informa-tion they have put in. They will want electronic submittal of applications and supporting documents to help reduce costs to themselves and their clients.

The most efficient way to determine what functionality is required for these users is to review existing applications they have completed using the present process and determine whether the system reproduces this process.

The use of the Internet or an in-office kiosk terminal can be of significant ben-efit to an agency in this process. It is often overlooked that if a data entry system is put online, the data entry process typically performed by the administrative staff is now virtually eliminated. Rather than entering the data themselves, now all the administrative staff needs to do is validate and verify that the data are complete and accurate. Furthermore, much of this validation and verification can be handled by the system itself. The use of pick-lists, fast matching, data validation, and other types of technology can have significant positive effects on the process of imple-menting a system.

Citizens

The final class of users is the citizens, or the general public. These are users that as individuals have an extremely low usage rate for a system such as this, but as a group, as a whole, they are significant consumers of the technology. They often have needs that are very focused and limited and very reproducible. For this user, if access is provided, a simple and logical interface is needed because these users may only use the system very infrequently.

Other Factors with Users

A number of other factors should also be taken into consideration when assessing the needs of the users, including their frequency of use, level of computer profi-ciency, and receptiveness to change.

Frequency of Use

The users of a GIS can be broken down into four major types based of the func-tion they will perform with respect to the system itself. These categories are described in the following list. Each defines the role that a user will play in the sys-tem and the level of expertise that they will need to have with the technology.

- *Data maintainer.* Maintains the data layers and associated attributes in the system. Because these people perform the data maintenance, they typically have a high level of expertise with the technology. Training for these users typically includes intensive technical coursework.

- *Data owner.* A user or department that is responsible for a data layer. In most cases this user is the same as a data maintainer, but in some cases a data owner may be a different user. An example is when an agency has a GIS coordinator who is responsible for making physical changes to all data layers, but other departments are actually responsible for what the content of the layer is.

- *Power user.* Use GIS on a regular basis to perform day-to-day functions. These people usually have a high level of technical understanding of GIS applications as they pertain to their department or their industry.

- *Occasional user.* Uses the system to perform basic review and query functions. This user typically just queries the system for basic information, does not typically perform advanced analysis with the system, and only has a basic understanding of GIS.

Computer Proficiency

Not only is it necessary to identify and classify the users of the system based on the functions they perform, it is also important to further classify the users as to their computer proficiency. It is very common for agencies to be planning an implementation of a GIS from what is presently a paper-based system. Staff members who may be presently performing the functions in the paper-based system may have little to no computer experience and may initially resist the change. Another common issue is that the present staff may be using a more text-based type system and may have little to no Microsoft Windows experience. It is important to identify these issues up front and provide proper basic training if necessary.

Applications

The next important step in the needs assessment process is to determine which functions within the business processes that are going to automated can benefit from GIS technology. The definition of an application can take on two primary forms with respect to GIS technology. In the first, GIS is the core technology used for the solution to provide spatial analysis. Typical examples of these types of applications are tools that use GIS to answer the following questions:

- What exists at or near a particular location?
- Where are all of the places where this exists?
- What geographic areas meet the following criteria?
- What has changed in this area or region over time?

- What spatial patterns exist?
- If we where to do this, what would the result be?

The following are a few examples of specific types of applications or functions that are solved by answering the above questions.

View, Locate, and Query Data

Viewing, locating, and querying data are the most generic of GIS applications. Almost everyone who uses GIS is interested in viewing some type of mapping information. The users need to perform functions such as finding a certain parcel, finding where all features of a certain type are, drawing features with different colors or symbols, or finding records in a database and then seeing where all of these features are on a map.

Complaint Tracking

A complaint-tracking application automates the process of taking a call or complaint from a concerned citizen or business owner in an agency, forwarding the call to the appropriate department or departments, and notifying, inspecting, citing violations, and tracking the actions taken in response to this call. Depending upon the department, complaints can come in many forms, and they often begin in the wrong department. The application should allow for any department's staff member to see any complaint and any actions taken by other departments. These complaints and actions can then be viewed in a mapped form to look for spatial patterns.

Mailing List Generation

Mailing list generation is more of a tool than an application. This tool should have the capability of creating a mailing list to be used to either create mailing labels in standard formats or to use in form letters. As an example, all properties within a certain distance of a subject property, a water line, or a roadway can be determined and their addresses fed into a form letter and used to create mailing labels. A customized mailing list can be created for the whole agency, for a select geographic area, or for properties, customers, or potential customers meeting certain criteria.

Address Matching (Pin Mapping)

Various databases and paper documents contain information relating to a property location. These database records and/or documents can be linked to a building or parcel by their street or mailing address. The result of this linking can be a point at the middle of a building or parcel. It may also result in a map with shaded buildings or parcels by which records match. If an agency lacks an accurate base map, low-cost street centerline files can be purchased that contain associated address ranges on the left and right sides of the roads, and these ranges can be used to geocode the points that you wish to map to perform basic spatial analysis.

Maintenance Tracking and Inventory

Maintenance is performed routinely on both the indoor and outdoor components of facilities and properties throughout an agency. During this process, many forms are filled out, creating an inventory of many attributes of the features that need to be tracked by various departments. Even without the forms, maintenance of properties can be tracked in a database linked to the property, provided that property is mapped. On occasion, descriptive information relating to these structures is changed due to repairs or replacements. This application allows the input of many characteristics of these features using on-screen form menus, and then analysis can be performed on this data to help schedule more effective maintenance programs, systematic repairs, or efficient total replacement.

Incident Mapping and Tracking

Often departments deal with situations that don't necessarily involve permitting or complaints but are related to an emergency or other unforeseen events. GIS can be used to zoom to where an incident occurs, point to where it happened, create a new point of where it happened, or add a record to the database that stores the building, property, street segment, or other GIS mapping feature. A form will display on the screen, allowing the user to enter information pertaining to the incident, along with the name of the caller, name of person taking the call, and the resolution. Emergency management services departments can then respond to the incident in the most efficient manner using a routing application, and once they arrive on the scene, they can locate the facilities they need to support the incident using the technology. Fire departments in urban areas use this type of application for response and need to know where the closest fire hydrants are when they get there. In rural areas the fire department may need to know where the closest pond or swimming pool is. GIS is also commonly used by public safety officers to analyze patterns of occurrences over time to help solve repetitive crimes or better plan and dispatch officers based on the past history of events.

Routing Applications

A routing application allows someone to plan the most efficient way to get to multiple locations in the most efficient manner. The simplest example of this is Web-based direction sites such as Mapquest.com that allow you to determine the most efficient way to get from one location to another. A more advanced example would be a routing package that allows a school system to plan the most efficient route to pick up all of the third and fourth graders and get them to school on time.

Permitting

Permitting processes and systems are carried out by many agencies, and the functions within these process can benefit greatly from enhancement with GIS. Some examples of how GIS can be used in this application are as follows:

- Check distance from proposed structure to other features such as property lines, setbacks, wetlands
- Map where all building permits have been issued this month, quarter, or year
- Map all places where a specific contractor is doing work in the agency
- Map where all inspections are for this week by inspector
- Calculate fastest route to get to each inspection
- Verify zoning of a parcel
- Generate mailing labels and abutter's notification letters for all parcels within 250 ft of a parcel in less than 1 minute
- Map all septic failures that have occurred this year
- Calculate distance to nearest day care center or residential zone from a specific parcel

The second definition of a GIS application is when GIS acts as the integration technology that helps store and retrieve information based on where it physically is located. One of the most difficult problems to solve in an integration project is identifying what common characteristic can be used to store and retrieve information in one or more systems. GIS allows the integrator to use the physical location of the feature as a method of storing and retrieving the appropriate piece of information. A few examples of systems that are in place in many government agencies where GIS can be beneficial as an integration technology are as follows:

- Computer aided mass appraisal (CAMA)
- Enterprise resource planning and financial (ERP)
- Permitting
- Document management or imaging
- Computer aided dispatch and records management (CAD/RMS)
- Computer aided design and drafting (CADD)
- Complaint tracking
- Customer relationship management (CRM)

Data are one of, if not the most, important component of a GIS system. Three primary types of data comprise a GIS: spatial data or layers, attribute data, and documents or images. Determining which of these items should be included in a system, how accurate they should be, how complete they need to be, what order they should be created in, and how they should be created are all of the questions that need to be answered. Let's try to answer each of these questions one at a time.

Evaluating Existing Data

The first decision that needs to be made is which layers need to be created to support the system. This is most easily accomplished by first inventorying what layers are in use in an agency. Table 2.1 is a typical list of layers that are used by a local government agency to perform their duties. Table 2.2 shows the 13 departments that were expected to have some GIS use in this community and the 33 existing paper and digital databases that contain geographic information. Due to space constraints we can show only a partial set of each in further tables. This entire list can be given to each department, and each department can then be asked to rank each of the data layers based on its frequency of use. For example, in a typical ranking system layers that are used multiple times a day could be given a factor of 10, layers used multiple times a week could be given a factor of 5, layers used once a week a factor of 3, layers used once a month a factor of 2, and those used only a couple of times a year, a factor of 1. All of these factors are then summed for all departments, and a total usage ranking can be calculated for each layer. The layers can then be sorted from top to bottom, from high to low rankings, where layers having the highest total ranking would be used more frequently by all departments, and layers having a lower factor would be used less frequently. The results of this analysis can formulate the priority order in which layers should be automated to provide the highest level of usage to the most users (see Table 2.3). The one problem with this type of analysis is that it does not take into account the value or relative importance that a layer can provide. For example, evacuation zones are a layer that is not used very frequently, and this analysis would confirm this, but when this type of a layer is needed, its importance can greatly outweigh its usage rate. Careful consideration has to be given to each and every layer to make sure that these types of factors are taken into account.

Table 2.1 Data Layer List

Features Layer (Point, Line, Polygon)	Description
Annotation	Miscellaneous text shown on a map
Rights of way (easements)	Boundaries of an area owned by one party with limited rights to others
Parcels	Boundaries of a contiguous piece of land in the possession of an owner
Topography (10-ft contours within regional watershed)	Lines representing a prescribed increment of elevation change
Topography (2-ft contours within town limits)	Lines representing a prescribed increment of elevation change
Building footprints	Perimeter of exterior of a building foundation
Sidewalks	Boundaries of sidewalks
Town line	Boundary of an area owned by a municipality

continued

Table 2.1 (Continued)	
Features Layer (Point, Line, Polygon)	*Description*
Wetlands	Boundary of an area designated as a wetland by a regulating agency
Walls	Stone walls, barbed wire fences, retaining walls, etc.
Zoning	Boundaries of zoning areas
Recreational areas	Athletic fields, golf courses, parks
Edge of driveway	Edge of paved or unpaved, gravel, or dirt driveways
Land use	Boundaries of land use areas
Trees vegetation (> 0.5 Acre)	Boundaries of trees and/or vegetation areas
Hydrographic Features	Canals, marshes, piers/docks, dams, lakes, reservoirs, rivers, streams, creeks
Edge of pavement	Edge of a paved roadway
Utility pole and towers	A structure that holds or supports a utility line
Census boundaries (demographics)	Boundaries of census tract or block groups
Ponds	Boundaries of stagnant water without or with a small outlet
Roadway intersections	Graphical point depicting where road centerlines intersect
Utilities, gas	Graphical representation of natural gas distribution system
Railroad	Railroad tracks or yards
Storm drainage system	Graphical representation of storm sewer collection system
Swimming pools	Graphical representation of public and private pools
Traffic features	Signals, streetlights, signs
Water supply system	Graphical representation of water distribution system
Drainage features	Floodwalls, headwalls
Flood zones	Boundaries of the edge of flood waters in varying flood frequencies
Protected open space (public & private)	Boundaries of protected open space areas

Table 2.1 *(Continued)*

Features Layer (Point, Line, Polygon)	Description
Sanitary sewer systems	Graphical representation of sanitary sewer collection system
Street centerlines	Graphical representation of the center of a street
Bridge	Graphical representation of the structure of a bridge
Channel encroachment	Boundaries of restricted areas along navigable waters
Miscellaneous structures	Above-ground tanks and other misc. structures
Groundwater recharge areas	Boundaries of a groundwater recharge areas
IRA map layer of police zones	Graphical representation of police zones
Residential wells	A hole or shaft in the ground used to provide drinking or irrigation water
Wellhead protection areas	Boundaries of protected areas around surface water wells
Groundwater protection overlay district	Boundaries of a groundwater protection overlay district areas
Soils	Boundaries of differentiating soil types
Surface drainage areas	Boundaries of areas that drain to storm drainage structures
Utilities, cable	Graphical representation of cable television distribution system
Utilities, electric	Graphical representation of electrical power distribution system
Curbing locations/types	Graphical representation of curbing material types
Groundwater aquifer	Boundaries of a ground water aquifer areas
Paved aprons	Driveway area between the front of sidewalk and edge of pavement
Plowing/sanding routes	Lines depicting routes followed by plows and sand trucks
Public transportation (bus routes)	Lines depicting regular routes that buses travel

continued

Table 2.1 (*Continued*)

Features Layer (Point, Line, Polygon)	Description
Town network cabling (fiber optic)	Graphical representation of town's fiber optic network
Municipal wells	Municipal drinking water wells
Pavement lines/striping	Graphical representation of roadway pavement striping and marking
Utilities, telephone	Graphical representation of telephone system network
Survey monuments	Physical monuments established townwide with known coordinate values
Voting district maps & lists	Boundaries of voting district areas
Trails, greenways, & bikeways	Lines depicting official recreational routes maintained by the town or state
Leaf vacuuming routes	Lines depicting routes followed by maintenance for leaf collection
School bus routes	Lines depicting routes followed by school buses for pick-up and delivery of students
School district maps	Graphical representation of school district boundaries
Street sweeping routes	Lines depicting routes followed for street sweeping
Trash pickup routes	Lines depicting routes followed for trash collection
Cemetery plot layouts	Areas depicting ownership of burial plots

Table 2.2 Departments and Databases

Departments	Existing Databases
Fire, Rescue, EMS	Assessor's CAMA
Engineering	Building permits
Town clerk	Outstanding fees/taxes
Highway & fleet	Building inspections
Planning & zoning	Sewer accounts
Water & sewer	Water accounts
Budget/citizen services	Restaurant licenses
Assessor	Hazardous waste storage
Information systems	Pavement management/conditions
Parks & cemetery	Septic permits
Police	Well permits
Building	Trade names
Health	Contractor licenses
	Call-before-you-dig requests
	Cemetery ownership
	Curbing locations/types
	Dog licenses
	HTE work orders
	Pavement lines & striping
	Sewer permits
	Water & sewer maintenance (work orders)
	Water cross-connection data
	Water permits
	Alarm registrations/key holder Info
	Computer inventory
	Food service licenses
	Wastewater quality data
	Water quality data
	Complaint tracking
	P&Z applications
	Special health needs
	Water source production data
	Permit holders (alcohol, game room, etc.)

Table 2.3 Department by Feature

Feature Layer Ranking	Features Layer (Point, Line, Polygon)	Description	Assessor's Source	P&Z Source	Engineering Source	Fire, Rescue, EMS	Engineering	Town Clerk	Highway & Fleet	Planning & Zoning	Water & Sewer
					Department Totals	15	15	12	12	15	15
					Department Ranking	1	2	3	4	5	6
1	**Annotation**	Miscellaneous text shown on a map	Engineering		Building As-builts	3	3	3	3	3	3
2	**Rights of way (easements)**	Boundaries of an area owned by one party with limited rights to others	Clerk	Engineering	Assessor	3	3	2	3	3	2
3	**Parcels**	Boundaries of a contiguous piece of land in the possession of an owner	Engineering	Assessor/ planning	Assessor	3	3	3	0	3	2
4	**Topography (10-ft contours within regional watershed)**	Lines representing a prescribed increment of elevation change	Engineering	Engineering	Engineering	3	3	2	3	3	3
5	**Topography (2-ft contours within Town limits)**	Lines representing a prescribed increment of elevation change	Engineering	Engineering	Engineering	3	3	2	3	3	3

Another factor that this analysis helps determine is which departments would have higher usage rates of the system. If you now total the rankings given to all layers by department and sort this ranking from high to low (left to right on this chart), you can determine the more frequent users of the system. Departments listed on the upper left portion of this matrix are those that will use the system the most, and those listed to the right will be less-frequent users.

Two other factors that should be inventoried as part of this analysis include which department is responsible for the layer and which department maintains the layer. Often agencies have a central department that performs mapping updates, whereas another department actually is responsible for the content of the map. For example, the assessor is usually the department that is responsible for the content of the tax maps an agency uses, but often another department, such as engineering, may perform the actual updates because engineers are proficient with drafting techniques or a CAD program. These relationships will be important to the design of the maintenance strategy of a GIS program.

The next factor that is documented is the source that will be used for the automation. Again, there are often multiple departments within an agency that may have different copies of the same type of mapping. The assessor may have tax maps that show features that are pertinent to the assessing function, and engineering or building may have another set for tracking house address assignments. It is important to document all versions of these maps and more importantly to document the differences between each so that critical features or attributes are not missed in the system.

The final factors that should be developed as part of this matrix are the methods that should be used to automate the layer and the frequency at which maintenance should be performed. The methods used for automation are discussed in chapter 7, Data Development and Conversion Plan, and maintenance frequencies are discussed later in this chapter.

The matrix in Table 2.4 is a summary of the same type of process used to analyze database systems that is in use in an agency and can be tied into using GIS. The same type of a ranking approach can be used to analyze the frequency of usage of typical databases in a GIS. Again, databases receiving the highest rankings will be used the most by the end users of the system, and the departments receiving the highest total rank will be the most frequent users of the system. Only a small set of the 33 database systems and the 13 departments are shown.

In addition to factors similar to layers such as source and methodology, there are a number of other critical aspects that need to be determined for a database. Connect to is the description of the spatial feature that the database would be connected to for retrieval. Many times there are multiple features that these attributes would be connected to for retrieval based on the application in which it will be used. Archive years is the duration or time period for which historic data need to be tracked. Some attributes are only as important in their present state, whereas for others, complete, historic data are important.

Table 2.4 Database Creation and Maintenance Needs

Database Ranking	Database	Responsible Department	Department Totals						Source	Methodology	Connect to	Archive to M5	Connection Type
			89	36	55	28	22	19					
			1	2	3	4	5	6					
			Water & Sewer	Fire-Rescue-EMS	Building	Budget Citizen Services	Health	Assessor					
1	Assessor's CAMA	Assessor	5	1	5	5	3	5	Assessor's vision system	Connection to new vision system with appropriate security rights	Parcel	All current data	Dynamic through SQL link
2	Building permits	Building	5	7	5	5	1	3	New permitting system	Connection to new permitting system with appropriate security rights	Parcel & structure	All electronic historic data	Dynamic through SQL or ODBC link
3	Outstanding fees/taxes	Collector	5	7	5	5	0	0	HTE financial system	Connection to HTE system with appropriate security rights	Parcel	All current data	Dynamic through SQL or ODBC link
4	Building inspections	Building	1	7	5	1	1	0	New permitting system	Connection to new permitting system with appropriate security rights	Parcel & structure	All current data	Dynamic through SQL or ODBC link
5	Sewer accounts	Water & sewer	5	0	5	5	2	1	Sewer billing records	Quarterly dump of sewer billing records	Parcel & structure	Current year	Static

The final factor that is important to a database is the type of connection that will be made to the spatial feature. Historically, the most common connections are static dumps of data where a snapshot in time is taken and the data are converted and linked to the spatial feature. In recent years it has become more common for databases to be tied into in a more live manner using either Structured Query Language (SQL) connections or open data base connectivity (ODBC).

Table 2.5 shows the final data analysis for documents or scanned images that could be tied into from the system. Just like the last two analyses, documents are inventoried and ranked based on usage, sorted, and comparatively ranked by their total anticipated usage. Departments are again ranked by their overall usage rankings and the sources of the images are documented. With documents, a few additional factors should also be determined. The total quantity of documents is important for estimating the cost or effort for conversion of the documents into the system, and go-back-years, similar to archive-years in databases are important to determine how critical the historic version of this document is.

Three major ways to handle historic data are typically found with documents. The first is that all historic documents, such as land records, are important and should be included. In this case all of the historic data need to be converted, indexed, and included in the system. The second is that only a certain period of time is critical for the users of the system, such as 3 to 5 years' worth of historic data. Finally, go forward means that historic data are not that important or may be too cost-prohibitive to convert, and only new documents or recently accessed documents, going forward, will be collected an input into the system.

Accuracy

The accuracy required for any GIS layer can also be another complex factor to calculate because as the accuracy of a data layer is increased, the cost or time it takes to automate a layer also goes up, usually not in a linear manner. The difficult decision is to determine what is the appropriate accuracy to develop to. It is very easy to say that the accuracy that a data layer should be developed to should equal the highest level of accuracy that any one department needs. If cost were not an issue, this would be the appropriate approach. The recommended approach is to perform a standard return-on-investment study and determine whether the higher level of accuracy justifies itself over a 5- to 10-year period. Accuracy is further discussed in the section in chapter 4 "Accuracy, Precision and Completeness."

Completeness

Completeness is the final factor that needs to be considered when performing an assessment on the data to be included in a GIS. Again, striving for 100 percent complete data, in an ideal world, is an easy goal to set but a very difficult one to achieve. Again, balance has to be considered to determine where the return on investment lies with respect to the usage rate of the data. In addition, it is also

Table 2.5 Scanned Images

Scanned Images Ranking	Scanned Images	Responsible Department	Water & Sewer	Assessor	Budget/ Citizen Services	Engineering	Planning & Zoning
		Department Totals	45	42	25	21	20
		Department Ranking	1	2	3	4	5
1	**Property photos**	Assessor	1	5	5	1	5
2	**As–builts (building)**	Engineering/building	1	5	0	5	3
3	**Land records (plans)**	Town clerk	4	5	5	5	5
4	**Land records (deeds)**	Town clerk	3	5	5	5	5
5	**Septic system as–builts**	Health	2	1	5	1	0
6	**Sewer laterals**	Water & sewer	5	1	5	1	0

important to determine what the potential liability would be if the data were less than 100 percent complete.

Maintenance

The maintenance of the data in a GIS is one of the most important aspects of ensuring continued success of the system, the mapping features being the most important part of the data. It is imperative that these features be updated on a regular basis to maintain the integrity and accuracy of the mapping in any system. Without proper maintenance, other departments that use the mapping in a GIS would soon lose confidence in the information provided by the mapping. The end result would be a product that no longer performs its function as a foundation for a GIS. The development of a maintenance plan after implementation is discussed later in the book

Frequency of Updates

Another important consideration for maintaining the data in the system is the frequency of updates for each of the data elements in the system. The most common definitions of the frequencies of typical update cycles and when they are used are as follows:

- *Semi-monthly.* This type of update is planned for data layers that have the highest level of importance in the system. These data layers change quite often, and their changes are considered to be critical to decisions that will be made with the system. A typical example of this is a parcel base. Attribute data such as change of ownership, change of definition (subdivision, lot splits, or combinations), and/or improvement in accuracy (new surveys being filed) are also critical. These layers are also the ones that the highest number of departments have use for or depend on.

- *Monthly.* These layers still have a high level of importance, but typically they do not change as frequently as the previous layers. Typically the features on these layers take longer to construct (such as building footprints) and require an official approval or as-built to be filed before an update can be completed or accepted.

- *Quarterly.* These layers are those that change on a much less-frequent basis than the previous two types. They typically require a formal approval and acceptance by a department or commission and acceptance before they can be updated. The higher level of frequency does not justify the cost of maintaining multiple layers of each type that are approved, pending, and in the application process.

- *Annual.* These layers are typically used for planning applications. Street sweeping, snow plowing, and school bus routes are examples of layers that require a considerable amount of planning before they are defined, but once they are defined, they do not change very often throughout the year.

♦ *As outside source updates.* These layers are created and maintained by an outside source. Once they are created, they are not often updated by these agencies; therefore there is no need to update them unless changed. Examples of these are FEMA flood plains, soil type definitions, and public transportation routes.

Software Selection

There are a number of factors or considerations that one should take into account when selecting a GIS software product.

System Components

The first and foremost factor that should drive the selection of a software product is the business processes that the product performs. Bells and whistles will often make you want a product, but the core functions that you perform are the ones that should be looked at first. A good rule of thumb is that if you look to automate the processes that you spend 80 percent of your time on, you will only spend 20 percent of the money that you would spend to automate the remaining 20 percent.

Ease of Use

The next most important factor is the ease of use of the product. This is twofold. Beginning users should be able to pick up the product and easily work through functions with the use of wizards, but expert users should be able to navigate through the product in a more efficient manner. The best products on the market have both novice and expert modes that allow you to switch over as your proficiency with the product increases.

If the product is modular in structure, each module should have the same look and feel so that once users learn the first module, they can make an easy transition to another module. The best way to fully understand how easy a product is to use is to talk to both new users and users who have been using it for a long time and consider themselves experts.

Product Support

Another important consideration for a product is how the parent company provides technical support. Is it something that is included with the product? Is it free for a period of time? Good technical support groups are made up of good qualified technical support people. Good people cost money, so be leery of free technical support.

Another consideration with support is the different avenues that are provided for support. Email, Web site, phone, and on-site support are all options. Be aware of the company's location, what time zone it is in, and what hours, in your time zone, it provides support. If you are trying to get that report out for a meeting that night, is support going to be available? How about that morning?

Training

Training is another important consideration. Who does the training? Does the company have dedicated training staff or do it send out sales people or developers? How much training comes with the software product? The most successful training programs are designed around a two-phased training program. The first phase is a complete course that introduces the users to all aspects of the product. After taking the course, the users return to work and are given phone support to work through any daily issues they come across. After 2 to 3 months the users receive a second course. This course is usually designed around specific questions developed by the users and supplemented with more advanced topics developed by the instructor.

Product Maintenance

The final consideration is product maintenance, which consists of the way the company handles bug fixes and new releases of the software product. Ask the company how often it puts out bug fixes, and how many it has put out in the last year. If it is too often, less than once a quarter, and the product is relatively new, there could be problems with the product. If the fixes seem far apart, only twice a year, and the users say the product is buggy, the company may not have enough developers on staff. It's a delicate balance.

Technical Environment

Operating System

Another consideration for any of the product solutions is the operating systems that they support. Virtually every software product supports 32-bit environments such as Windows NT, 2000, or XP, but some of the systems do not support older 16-bit environments such as Windows 95 or 98. This can be an important consideration if you are using older hardware and you don't have a replacement budget. Web-based solutions are becoming more and more popular these days because the majority of the applications that can developed using Web technology will function independent of the end-users operating system. Using this type of an approach, 16- or 32-bit systems are no longer a barrier, and applications can be developed and used on many older computers without having to invest significant amounts of money for new infrastructure.

Network Environment

A network environment is an essential component for any GIS solution today. The question isn't whether or not there should be a network; the main consideration in this area should be the speed of the network that the system requires. Systems based on MS-Access technology require, at a minimum, 100 baseT network speeds, whereas systems that are Web based or use a client server (Oracle or MS-SQL Server) can run very efficiently on 10 baseT networks. For a smaller agency, a MS-Access system can be successfully implemented on the slower network, but for larger municipalities, a MS-Access solution can only be effectively deployed if technology like Citrix or Terminal Server is used.

The end of a needs assessment is the time to select software and examine other technical requirements; at the beginning you may not know all your needs, and certain software is better at supporting certain needs. Software selection and some of the other technical issues of implementation are discussed further in a later section

Database Considerations

The final area that should be considered is the back-end database that is used with the system. The most commonly used back ends are MS-Access, MS-SQL Server, and Oracle. MS-C.V. Access-based systems require the minimum amount of expertise at the end-user site, and need for expertise increases as you transcend to SQL-Server and then to Oracle. Care should be taken to appropriately size the system based on the number of users, the number of transactions, and the level of expertise at the end-user site. Database selection may be constrained by existing databases in use by the organization, and you should certainly select your database software before you begin the design phase for the GIS data because there are design considerations specific to different database software systems.

Assessing Costs and Benefits

The costs and benefits of implementing a GIS system has been one of the most widely asked and debated questions when planning the implementation of a system. This question has been a difficult one to answer because of the relative newness of the technology and the lack of concrete facts and statistics on which to base the analysis. Luckily, in recent years more concrete data have begun to emerge that can be applied to some of the applications of the technology. The following discussion describes the process that one can go through to perform this type of analysis and concludes with some facts and figures that have been found that may be useful to an organization in this process.

Predicting the costs for implementing a system is a relatively straightforward process. Because many systems have been implemented to date, an abundance of information is available about the cost of building system components. There are two primary breakdowns to the costs that are associated with implementing a system: initial and recurring costs. Fundamental initial costs are easily identifiable and are composed of the initial capital outlays that must be made to plan the system, develop the data to support the system, purchase the hardware and software necessary to support the system, and provide initial training to the users. There is an important fact that is often overlooked when identifying these costs, especially in the data area. Statistics have shown that approximately 80 percent of the cost of implementing a system are typically accounted for by the cost to develop the data in the system. Often this data would have been developed whether or not a GIS was implemented. For example, traditionally many government agencies would develop base mapping for capital improvement projects they are performing to use for engineering design or survey and construction documents. These costs would be incurred regardless of whether the data are developed in a GIS format. For the purposes of the cost and benefits analysis it is important to identify only the difference in cost that is incurred because the data are developed for use in a GIS, not the total cost to develop the data. Another example is tax mapping. Parcel base creation is one of the higher cost in a systems implementation, yet often tax mapping and tax mapping maintenance is already being performed in the agency. Again, for the purposes of the analysis, only the difference in cost should be accounted for, not necessarily the total cost.

Recurring costs are those that are incurred repetitively over time such as maintenance costs, staff salaries, possibly some data costs, and training costs. The same rule applies to these costs. If maintenance of a data layer was already occurring, only part of this cost, the increase due to GIS, should be accounted for in the analysis. The addition of a new staff person such as a GIS coordinator should be accounted for in its entirety, but if existing staff who were already performing data maintenance functions, are retrained to perform these same functions in the GIS, their costs are offset, not additive. Often, once created, efficiency is gained by implementing the system, and the actual cost to maintain the data in a centralized location can reduce the effort to maintain the data in the future. Therefore, the maintenance of a GIS can actual be a positive return, not a cost.

A very basic, yet fundamental example of this is an owner name change. A study was done in a municipal agency recently where the amount of time that was spent changing records in computer programs, databases, and hard-copy documents every time a new owner moved into town or changed a name was accounted for. The municipality had a population of approximately 21,000 people, and it was estimated that 3,000 individual changes to names took place each year. There are 26 departments in this municipality, and prior to implementing system, 20 people in these departments made a change in some document, spreadsheet, or database. Each change took approximately 5 minutes to make on average. If you do the math, one owner change accounted for 100 minutes of staff time, and in all a total 5,000 staff hours, or 2.8 staff years, were spent changing names. The design was to implement a system where one person now changed the name, and others used this name for their purpose from a centralized GIS. The one person now only spends 250 hours annually making this change, and 4,750 hours of staff time is gained on an annual basis. This community on average has a high rate of name changes, but experience has shown that virtually every agency has this type of redundancy in many of its systems including accounting, permitting, and taxation.

This leads into a discussion on the typical benefits that can be gained from implementing a system such as a GIS. Two major classifications of benefits can be achieved from implementing a system: quantitative and qualitative. Quantitative benefits are those like the previous one in which an exact quantity of time, effort, or cost can be derived from the analysis. Some other examples of quantitative benefits are as follows.

A medium-size municipality in northeastern Massachusetts implemented a parcel-based GIS as part of the revaluation process. When a link was made between its CAMA system and the tax maps, a number of inaccuracies in the data in their CAMA system were found, including properties that were not being taxed, properties with significant errors in the land area they were being taxed on, and physical features such as sheds and pools that were not accounted for because the owner had built them without a permit. The GIS was used to analyze and compare the mapping data to the assessment data and $200,000 a year in additional tax revenue was found due to the application of GIS.

Another medium-size municipality in the greater Boston area has implemented a Web-based GIS system to give citizens, attorneys, and real estate professionals access to data such as owner and parcel information, tax assessment information, comparable sales data, and basic mapping. After implementing the system, the municipality had a significant decrease in counter traffic, reducing staff time by 400 to 500 hours per year due to the lack of need of answering questions and processing walk-up requests.

A relatively small municipality in central Connecticut, population 12,500, recently implemented a GIS system to automate its sewer, water, and parcel information. One of the applications developed was designed to link the water and sewer account billing system to the parcel base. Once linked a comparison was run between accounts that have received sewer and water assessment charges that were within the sewer and water district and accounts that were also being billed user fees. A significant number of properties were found that were not ever billed assessment fees, but should have been, as well as a significant number of accounts that where not being billed annual fees, yet should be. The end result of the project was that a one-time revenue of $26,000 was realized from assessment fees, and an increase of approximately $6,500 per year was realized from annual service fees.

To perform a cost-benefits analysis, one of the key factors in applying some of this data is to make sure that the data are all standardized to a consistent unit of time. For example, recurring costs and benefits, by definition, repeat themselves on a year-to-year basis. Initial costs are a one-time expense. If the life of a one-time cost covers a 5-year period, this cost must be spread out over the 5-year period even if it must be paid for over 2 or 3 years. In addition the costs of some of the initial expenditures, like data development and hardware and software purchases, often still have a valley at the end of their life period, so standard depreciation calculations should be applied to the component because it is an asset of the agency.

To summarize, quantitative benefits are those that are measurable or quantifiable. Typical quantitative benefits are reductions in staff time to perform a task, decreased operating costs, increased revenue, cost avoidances, and cost reductions.

The second category of benefits that can be seen from implementing a GIS is qualitative. Qualitative benefits allow people to make better decisions, decrease uncertainty with an issue, or improve the image or services that are provided. These benefits are extremely difficult, if not impossible, to put a value on. No one would argue that having better information available to make a decision is not a benefit, but the only way to quantify that benefit is to first not have the information available, make a mistake, and then have to in some way pay for that mistake. Providing better public service is also considered a benefit to some, but not all, so putting a measurable value on it is difficult. Other examples of both quantitative and qualitative benefits include the following:

- ◆ Effort to produce maps by manual means was greater than total cost of making identical maps using GIS: quantitative benefit
- ◆ Use of GIS allows garbage and recycling collection efforts to be reduced through better scheduling of collection routes: quantitative benefit
- ◆ Emergency vehicles reduce average arrival time by using GIS: both
- ◆ Information from GIS was used to avoid lawsuit in land ownership case: qualitative benefit
- ◆ Board of Education finds a better location for a new school through use of GIS: qualitative benefit

Pulling the Needs Together

The final stage of a needs assessment is to pull all of the components of the system together into a phased approach and develop an implementation plan. The needs assessment portion of the project can be considered the wish list for an implementation, and the implementation plan can be considered the do list. The strategic implementation plan includes a number of fundamental components such as the following:

- ◆ Data development plan
- ◆ Application development plan
- ◆ Staffing and management plan
- ◆ Training plan
- ◆ Hardware and software acquisition plan
- ◆ Financial projection plan
- ◆ Schedule

As described, the needs assessment constitutes a wish list, meaning that all the identified needs are addressed, regardless of cost or priority. Implementation planning attaches costs and priorities to the needs. Factors that affect the costs and priorities need to be analyzed, available budgets and time constraints must be considered, and a realistic strategy must be developed.

The following charts summarize the priorities of applications, data, and departments in the system design. Table 2.6 summarizes which departments will use which applications and what type of a solution will need to be developed for each application. Only 5 of the 13 departments are shown.

Table 2.7 summarizes what data layers will be required to support each of the applications that will be developed. Only 10 of the 60 data layers are shown. Table 2.8 depicts which databases (10 of 30 shown) will be required for each of the applications, and Table 2.9 depicts which of the imaging sources will be used by each of the applications. These summary tables combined with the previous matrices are the complete summary of what applications need to be done first; what spatial data, attribute data, and imaging data are required for each; what methods should be used to produce the desired results; and which departments should be included first. These combined with a detailed schedule that includes dependencies between the applications, data sets, training dates, staff additions, and hardware and software acquisitions constitute the complete implementation strategy. The only remaining piece is the budget, which is a derivative of the above documents.

Although the final needs assessment will contain quite a bit of text, these tables are the core of a needs assessment. The example followed was for local government, but the process fits any multidepartment organization with several applications and databases that require geographic information.

ADDITIONAL READING

In books and journals there is little material on how to conduct needs assessments; there are many statements about how important they are, however. Accordingly, the additional resources we suggest here are Web links and, as with all Web links, we can make no guarantee that they will still be there when you need more information. But a Web search on a good search engine should regenerate a similar list at any time with not too much effort.

INTERNET RESOURCES

Usually has links to recent needs assessments conducted by different organizations, guidelines for developing RFPs, and so on:
tenlinks.com/MapGIS/reference/best_practice/needsassess.htm

More guidelines for developing a needs assessment RFP:
archives.nysed.gov/services/local-gov/RFP.htm

The New York State Archives and Records Administration working with the National Center for Geographic Information produced a series of documents to help local governments in New York state implement GIS:
archives.nysed.gov/pubs/gis/needs.htm

The Nature Conservancy's needs assessment workbook. This organization has a Web-based enterprise GIS program as well:
gis.tnc.org/knowledge/capacity/needs_wkb.php

We have chosen not to list completed needs assessments of organizations that have placed them on the Web for fear that they will not stay there long. Nor have we included the many examples of needs assessments that consulting firms have placed on the Web as demonstrations of the projects they do. Before the World Wide Web people circulated copies of needs assessments through personal networks, but today there is no shortage of samples you can read for ideas on how to conduct one.

Table 2.6 Departmental Application Needs

		DEPARTMENT					
GIS Application Needs	Application Ranking	Water and Sewer	Engineering	Fire-Rescue-EMS	Assessor	Information Systems	Proposed Solution
Departmental ranking		10	9	9	8	4	
GIS view, locate, and query data	13	X	X	X	X	X	Web GOS and desktop GIS software
Complaint tracking	13	X	X	X	X	X	Land management software
Mailing list generation	12	X	X	X	X		Web GIS
Integration with existing applications	11	X	X	X	X	X	Custom solution
Address matching (pin mapping)	10	X	X	X	X		Web and desktop GIS viewer
Permit tracking	9	X	X	X	X		Land management software
Inspection scheduling and tracking	8	X	X	X			Land management software
Maintenance tracking and inventory	7	X	X	X	X	X	Asset Management
Incident mapping and tracking	5	X		X	X		Desktop GIS software
GIS data maintenance	2	X	X				Full GIS Development Software

Phase I

Phase II

Table 2.7 Data Needs

DATA LAYERS

	Annotation	Rights of way (Easements)	Parcels	Topography (10 ft. Contours)	Topography (2-ft Contours)	Building Footprints	Sidewalks	Town Line	Wetlands	Wells
Feature layer rankings	28	28	27	24	24	23	21	21	21	20
GIS view, locate, and query data	X	X	X	X	X	X	X	X	X	X
Complaint tracking	X	X	X		X	X	X	X	X	X
Mailing list generation	X	X	X		X	X	X	X		
Integration with existing applications			X							
Address matching (pin mapping)	X	X	X		X	X	X	X	X	
Permit tracking	X	X	X			X	X	X	X	
Inspection scheduling and tracking			X			X	X	X	X	
Maintenance tracking and inventory	X	X	X			X	X	X		
Incident mapping and tracking	X	X	X			X	X	X	X	X
GIS data maintenance	X	X	X	X	X	X	X	X	X	X

Phase I

Phase II

Table 2.8 Application Database Needs

DATABASES

	Application Ranking	Assessor's CAMA	Building Permits	Outstanding Fees/Taxes	Building Inspections	Sewer Accounts	Water Accounts	Restaurant Licenses	Hazardous Waste Storage	Pavement Management/ Conditions	Septic Permits
Database ranking		40	32	24	19	19	18	17	16	10	10
GIS view, locate, and query data	13	X	X	X	X	X	X	X	X	X	X
Complaint tracking	13	X	X			X	X	X	X	X	X
Mailing list generation	12	X	X	X		X	X	X	X		X
Integration with existing applications	11	X	X	X	X	X	X	X			X
Address matching (pin mapping)	10	X	X	X	X	X	X	X	X		X
Permit tracking	9	X	X	X	X			X			X
Inspection scheduling and tracking	8	X	X	X	X			X			X
Maintenance tracking and inventory	7					X	X	X		X	X
Incident mapping and tracking	5	X	X						X		X
GIS data maintenance	2							X			X

Phase I

Phase II

45

Table 2.9 Image Needs

IMAGES

	Application Ranking	Property Photos	As-Builts (Buildings)	Land Records (Plans)	Land Records (Deeds)	Septic System Asbuilts	Sewer Laterals	Town Facilities	Construction Plans	Field Cards	M45 (Property Assessment Forms)
Image ranking											
GIS view, locate, and query data	13	X	X	X	X	X	X	X	X	X	X
Complaint tracking	13	X	X	X	X	X	X	X			
Mailing list generation	12										
Integration with existing applications	11	X	X	X	X					X	X
Address matching (pin mapping)	10	X	X	X	X	X	X		X		
Permit tracking	9	X	X	X		X	X		X		
Inspection scheduling and tracking	8	X	X	X		X	X		X		
Maintenance tracking and inventory	7						X	X	X		
Incident mapping and tracking	5	X	X								
GIS data maintenance	2	X	X	X	X	X	X	X	X		

Phase I
Phase II

Designing the GIS Database Schema

D atabase designers use the word *schema* to refer to the diagram and documents that lay out the structure of the database and the relationships that exist between elements of the database. A schema is like a blueprint for a database that tells a knowledgeable builder exactly how to construct it. Naturally, designers spend a lot of time thinking about the schema. This work comes before worrying too much about the exact content of tables and even before design concerns for the spatial data. Rushing into building a database without laying out your schema is like trying to build a house without a set of plans; it might stand up for a while, but it will not be as useful as it could be.

The tools that assist in the construction of these schema are called computer assisted software engineering (CASE) tools. These same tools are used to design the structure of complex computer programs as well as databases, and most programmers know how to use them. Many in the GIS world do not, but it is usually possible to design your database with paper and pencil, and some database designers still work this way. The ability to erase entire tables, delete relationships, add relationships, and so on, is sometimes easier with pencil and paper or on a whiteboard than mastering a new set of tools. One of the problems with GIS is that it appears to force you to develop areas of specialization and skill that you didn't have before. Sometimes it just takes too long to learn the new tools, so feel free to use simpler ones you have mastered instead of new tools that do basically the same thing.

Elements of a Schema

A schema at its simplest consists of an arrangement of tables and the relationships between them. Because organizations differ so widely in the kind of work they do and the types of data they need to do this work, it is impossible to provide a cookbook schema for every application. Software vendors that have many users of a particular sort, however, have constructed template database schema that can be customized. But even the task of customizing an existing schema in a complex organization that is planning to construct a large geographic database that integrates most of its existing data and incorporating new data tables into that schema ought to be done by highly skilled database administrators and designers. From the perspective of the users of the geographic data, who may be a minority of the total set of users, it is important that your geographic feature tables be correctly linked to the other tables you need to do your jobs. Schemas for organizations such as electric or gas utilities can take up many square feet of paper and hundreds of pages of documentation, and it is not our intention to outline that process or run through an example like that here. Rather, our focus is to ensure that you know what a schema is, why its construction is important, and how to provide input and feedback to those who are designing the entire system.

Designers of schema may be strict or loose constructionists. A strict constructionist designer would insist on a larger number of elements in a schema and a correspondingly longer time to develop one. A loose constructionist would prefer a smaller but adaptable framework before beginning to assemble data, assuming that the schema will evolve over time and use. Here we take the second approach for the practical reason that schemas can get very large and involved and thus are difficult to discuss in this format and because we do basically feel that trying to tie down every possibility in a database reaches the point of diminishing returns quickly. So as a compromise Table 3.1 shows what we consider required and optional elements for a GIS database schema.

Data Dictionary

As a relatively simple example, we present a project conducted by a consulting company for an annotated bibliography of study reports and historical documents of the Snohomish River basin in the state of Washington. The requesting state transportation agency wanted a GIS interface on this bibliography and document collection so that by identifying a feature of interest, say, a particular wetland, users could pull up all the documents related to that feature and any recommendations that had been made for that feature with respect to mitigation, restoration, water quality, and so on. The data dictionary is shown in Table 3.2. This is a minimal but adequate example; it names fields with reasonable and brief names, and it provides information about the type of data the field contains, how many columns in the table have been allotted for a field, and a description of the field.

Strict constructionists might wonder why a Char (text) data type was given to the year instead of a numeric data type and what values the Rating field in the Citations table would be allowed to take or, in the Recommendations table Preserve field, what it would mean if a value was something other than 3. These concerns can be met in the metadata. The three tables that contain information clearly represent distinctly different objects or features. In the feature tables you find the geography, or the *where,* of the particular features, wetlands, environmentally sensitive areas, and so on. We are seeing the attribute table; there is an associated table that contains the spatial information that defines the features. There will

Table 3.1 Important and Optional Elements of a GIS Database Schema

Element	Description	Example
Important elements:		
Data dictionary	A field-by-field description of each field in each table. At a minimum it must include the data type (e.g., numeric, text, date, image), the spaces it requires in the field (if appropriate for the data type), and a description of the data. It should also include explanations of the various values that a record can take for each field. (See Data_Dict Table.)	Table_Data_Dict
Primary and foreign keys	Each table must have a primary key, a field containing an identifier for each record that is unique to that record. Foreign keys are fields in one table that are primary keys in another. Primary and foreign keys are used to link tables.	FEAT_ID in the Feature Attribute tables and table DOCREC_NO in the Recommendations (see Figure 3.1)
Entity-relationship diagram	A diagram using a standard diagramming format that shows exactly how the tables are related to each other – the primary and foreign keys and the type of relationship. It must show whether relationships are one to one, one to many, or many to many and what relationships are mandatory and which optional.	Figure 3.1, 3.3
Required metadata elements	See Metadata section later in chapter.	
Optional elements:		
Work flow diagrams	A diagram of exactly how various tables are incorporated into specific routine tasks through forms, reports, maps, etc.	
Form and report designs	The layout and data sources for normal data input forms and the design and data sources for standard output reports. In a GIS database the standard reports will include map or layout templates.	
Security	Explanation of how access to schema elements is allowed or restricted to users or classes of users.	
Domains and validation rules	As part of a data dictionary, explicitly defined allowed value ranges for data and acceptable unique values for certain fields. For example, the date allowed in the field for the date a permit process began could be restricted to be only the date on which the form was filled out.	
Standard queries	Regular queries that would access tables and be input to standard reports	
Optional metadata elements	See Metadata section later in chapter.	

Table 3.2 Data Dictionary

Feature Attribute Tables (One for Each Data Type, Point, Line, or Polygon). Contains only geographic information about the features. There will be a record for each feature.

Field Name	Type	Size	Description
GIS FIELDS	Char	Varies	Standard fields such as length or area
FEAT_ID	Char	5	Arbitrary and unique feature ID (e.g., PL001, LN001, PT001)
LOCATION	Char	80	Text description of the feature location (e.g., Sunday Creek)

Citations Table. Contains information about each document regardless of how many features or recommendations may be referenced in the document.

Field Name	Type	Size	Description
DOC_NO	Char	4	Arbitrary document number (D001, D002, etc.)
AUTHOR	Char	80	Author(s) written in citation format
YEAR	Char	4	Year of publication
TITLE	Char		Document title in citation (sentence) format
SOURCE	Char	100	Document publisher/source in citation format
RATING	Char	5	Document rating (e.g, b3 r1)
ANNOFILE	Char	12	Pointer to disk (text) file containing full annotation

Information about each recommendation. There could be many recommendations in each document, so DOCREC_NO is a unique number identifying each recommendation and its associated document.

Field Name	Type	Size	Description
DOC_NO	Char	4	Document number (see above)
REC_NO	Char	3	Sequential number for recommendations (R01, R02, etc.)
DOCREC_NO	Char	7	Combination of document and recommendation number
PRESERVE	Char	1	3 if recommendation is for preservation
MITIGATE	Char	1	Check box if recommendation is for mitigation
RESTORE	Char	1	Check box if recommendation is for restoration
FISHHAB	Char	1	Check box if recommendation pertains to fish habitat
WATERQUAL	Char	1	Check box if recommendation pertains to water quality
WETLAND	Char	1	Check box if recommendation pertains to wetlands
FLOODING	Char	1	Check box if recommendation pertains to flood control.
DESCRIPT	Char	200	Summary of the recommendation.

Link Table. Contains all the associations between map features and recommendations.

Field Name	Type	Size	Description
FEAT_ID	Char	5	Arbitrary feature ID (see above)
DOCREC_NO			Combined document and recommendation number (see above)

Source: Konkel 1999. Used with permission.

50

be several feature tables in this database, probably one for each class of features that is involved (a set of wetland features, perhaps a set of environmentally sensitive areas, etc.). The citation table contains information about the document itself; this table closely resembles a catalogue card.

The Recommendations table comes out of the needs of the project. The clients particularly wanted to track what recommendations had been made for the various geographic features. This meant that someone had to read each document and identify the various recommendations made for which features. So a recommendation is an object separate from features or documents. This is a key issue in schema design; each table should represent a distinct class of objects and only information that relates to that type of object should be present in the table. So a document object and a recommendation object represent different feature classes and belong in different tables.

It is very easy to fall into the practice of mixing feature types, that is, putting recommendation fields in the Feature table. This quickly leads to the question of how many recommendation fields to create. You can be certain that if you design the table with space for six recommendations, you will find a feature that has had seven recommendations made for it. Then you are stuck. Generally, if you need to decide how many fields to leave for a type of information, you have mixed feature classes in a single table. Consider another example of telephone numbers and people. A telephone number is not the same thing as a person. One number can be linked to many people and certainly one person can have many telephone numbers, but you really can't know how many numbers each person will have. So a well-designed database would have a separate table for people and one for telephone numbers with links between those tables. The telephone number table could also contain a field that identified the type of number, home telephone, work phone, mobile phone, and so on. Instead, what you usually see in a personal contact table is a separate field for each possible type of telephone number. But if a person has only one type of number, you are creating a space, but there is no data. This is another clue that you are mixing feature types in a single table.

Tables and Relationships

The second critical part of a database schema, and actually the one you create first, is a diagram that shows the relationships among the various tables in the database, as shown in Figure 3.1. Relationships have a property called cardinality that describes the type of relationship. The possibilities for relationships are one to one, one to many, and many to many. Additionally, relationships may have the property of being required (mandatory) or optional. An example of a required one-to-one relationship in this figure is the relationship between the Recommendations and Citation tables (in that direction). Each recommendation must have a document number (i.e., come from a document), and that DOC_NO links to the Citation table that contains information about the document. This means that a recommendation without a document is not possible in this database. If you want to allow for that, you can, but as this database is designed, all recommendations must exist in a document.

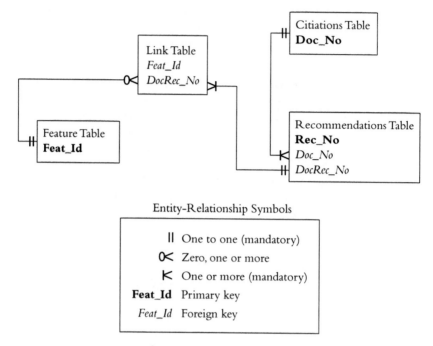

Figure 3.1   Schema diagram.

If a similar recommendation is made in another document, that is considered a different recommendation; otherwise the Recommendation-to-Citation relationship would be one to many. The relationship the other way (the Citation table to the Recommendation table) is a one-to-many mandatory relationship. Each citation must be related to at least one record in the Recommendations table but could be related to many if there were more than one recommendation in the document. Forcing at least one relationship means that all documents must contain at least one recommendation. If that were not the case, the relationship could be optional (i.e., 0, 1, or many and nonmandatory), but then there would be no link between the document (Citation table) and any features (real world geographic entities). It is only through the Recommendations table that you can tie geographic features to documents. The central relationship being modeled is between the geographic features and recommendations that have been made concerning them, and that is an optional many-to-many relationship.

To handle many-to-many relationships the schema needs a composite, or linking, table. Database programs relate tables one to one and one to many directly between the tables, but a many-to-many relationship requires the construction of an intermediate table. By establishing one-to-many relationships between each data table (the Feature and Recommendations tables) and the link or composite table, you create a many-to-many relationship between the two data tables. The one-to-many relationship between the Feature table and Link table allows features to exist in the Feature table that have not been discussed in a document and for which there are no recommendations (i.e., the feature may exist in the feature table and be seen on the map, but its Feat_Id value is not in the Link table). This means that geographic features that might be important for identifying where you are are not linked to any recommendations. Another way to handle this would be to create a dummy document and dummy recommendation that said "No recommendation has been made on this feature," possibly in the DESCRIPT field of the Recommendations table. Then the Feature table-to-Link table relationship would be a one-or-many (mandatory) relationship. You have the ability to design it either way. If a zero relationship is possible, clicking on the feature in the data (map) view will produce nothing. If you have created the dummy recommendation and citation records, the text "No recommendation has been made" would appear. If you plan it one way and change your mind, it is always possible to modify the schema, but it is better to think through questions like that at the beginning of the design process.

The reason that a schema diagram is important and not an optional element in designing a GIS should be clear from the preceding paragraphs. It is possible to document the relationships with words and descriptions, but the graphic picture of how the relationships flow is much clearer once you understand the symbols. With the tables and relationships in this schema it would be possible to click on a feature on the computer screen — a point (well), line (section of stream), or polygon (wetland) — and immediately know at least all of the following:

- Who made this recommendation and when was it made. (Table:Citation/Field:Author and Table:Citation/Field:Year)
- If this feature has had any water quality recommendations made on it and when. (Table:Recommendations/Field:Waterqual and Table:Citation/Field:Year)
- If this feature is recommended for preservation in any document. (Table:Recommendations/Field:Preserve)

Of more interest are the queries that this structure makes possible. For example, you could create a query that would show all features:

- For which a recommendation related to fish habitat was made between 1990 and 1995
- For which recommendations were made in a particular document
- That have a recommendation pertaining to wetlands and are recommended for preservation
- That have conflicting recommendations made for them in different documents

Schema Example

The schema example shown here is relatively simple. More complex databases may have dozens of tables and relationships. The development of a complete data dictionary and basic schema diagram before trying to populate it with data are vital steps. The steps to create these schemas are pretty straightforward. As an exercise we are going to work the process through for a local tax collector who wants a GIS database to collect taxes on four classes of features, land parcels, buildings, vehicles, and equipment. The key report that must be created by this database is a bill that will be mailed to an owner or owners so that taxes can be collected. The geographic features that this database must deal with are the land parcels (polygons) and buildings (polygons). The other feature classes, vehicles and equipment, have no inherent geography in this example and exist only as attribute tables.

Step 1. Identify all the possible classes of objects. We began with five object classes, land parcels, vehicles, equipment, buildings, and owners. But because this database is going to support a billing process, bills are another object class. It is important to separate the object classes so that you can design the appropriate fields for each class. A common design mistake in databases for land value assessment is to include the owner in the table for the land parcel. Owners and parcels are quite different classes of objects. They both have addresses, for example, but often not the same address. If you include the owner information as fields in the land parcel table, you run into the following problems:

- Some land parcels have multiple owners. You can allow for this with additional owner fields in the table, but most of them will be empty, and if you allow for only three owners, it is almost a given that you will find a parcel owned by six people.

- You have to store the owner's name for each parcel that he or she owns. The fundamental relationship between land parcels and owners is a many-to-many relationship. A single parcel may have more than one owner and a single owner may own more than one parcel. Storing owner information multiple times provides more opportunities for mistakes. Each owner will have only one entry in the owner table, and this removes the confusion between William J Smith, William J. Smith, and William James Smith. Perhaps this individual owns several parcels and is on the separate deeds with these slight variations on his name, but they all are the same person. If the name, rather than a parcel identification number, is part of the parcel table, it will be in the parcel table in these three slightly different variants.

- When an owner moves and changes addresses, you will have to make that change for every parcel that person owns. In a correctly designed schema each owner will be a single record in the owner table, and you will make the change once in that table. The owner's address is a property of the owner, not the parcel. If the parcel has an address, it is appropriate to have that in the parcel table, though. Addresses, although they may seem straightforward, can be rather complex things (see chapter 5 "How They Did It – Kansas Geospatial Data Addressing Standard."

Step 2. Sketch the relationships you will need between tables. At first, don't worry about the type of relationship, one to one, one to many, or many to many; just draw the minimum amount of needed lines. Figure 3.2 shows the six object classes and the relationships between pairs of tables.

Owners need to get bills and bills need the information from the land, vehicle, equipment, and buildings table. The relationship between the owner and the assets is one of ownership; the owner owns the asset. Viewed the other way, the asset is owned by the owner. The relationship from owner to bill is one of must pay and from bill to owner is must receive from. There are some relationships that might initially seem necessary (e.g., relationships between a bill table and the asset tables); however, they are not because they exist through the owner table. An additional relationship between buildings and land is shown in Figure 3.2. You might want to establish a relationship between the equipment and building and/or land table (is found in/on).

Step 3. Detail the key relationships first and then the secondary relationships. In this example you will need to decide whether you are going to send each owner one bill for all assets or a separate bill for each category of asset he or she owns. This is a decision that is independent of the database design; it can support either decision, but the design will differ. Making a decision to send out a single bill and later changing your mind to send a different bill for each category of asset would mean a redesign of the database. In our example, because the purpose is collecting taxes, the key relationship is between owners and bills. We have decided that we want owners who have any assets to get a single bill for all assets, but we also want the ability to include owners who own no assets at this time in the owner table. Perhaps they used to own assets but don't any longer. But if they own any assets, they are to get a single bill for all assets at once.

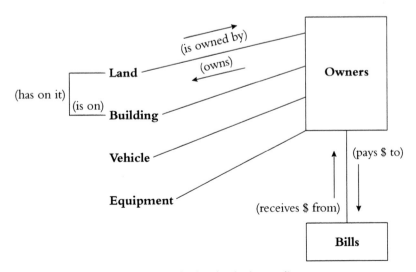

Figure 3.2 Early sketch of schema diagram.

Detailing the relationships forces you to think about primary and foreign keys for the tables. A primary key in a table is a field that contains a unique identifier that is not duplicated for any feature in the table. So, each owner has a unique ID number that might be a social security number, although there are occasionally duplicates of these numbers. Or your system could automatically create an integer Owner_Id value as owners are entered into the table. Most databases will automatically create a primary key for you, usually a sequential integer, but you may wish to create a key that contains information as well. A common primary key that could be created for the land table, for example, would be the combination of the map, block, and lot numbers of the parcel from the original paper maps. Foreign keys are fields that are primary keys in other tables. The relationships between tables are usually created between the primary key of the origin table and that field as a foreign key in destination table. For this reason keys must be formatted in exactly the same way. So if your primary key in the land parcel tables is Map-block-lot, the field must be built the same way in the asset table. If you stored it as Map/block/lot, the two tables would never join.

When joining tables, it is important to remember which is the from, or origin, table and which is the to, or destination. In Figure 3.3 consider the Owners table as the from table and the Bills table as the to table. The relationship from Owners to Bills is a one-to-many relationship because owners are going to be sent many bills through this system. Even though they will get one bill for all assets, this database will support many billing periods. Having no bills is a possibility because we wish to include people who are currently not owners but may become owners. To join these two tables the Own_Id, the primary key in the Owners table, exists as a foreign key in the Bills table. Looking at the relationship the other way, considering the Bills table as the from table and the Owners table as the to table, the relationship is a mandatory one-to-one relationship. Each bill must belong to one and only one owner. If an asset has multiple owners, only one receives the bill. This means that the Owner_Id values in the Bill_Table must be the ID values only for the primary owner. The multiple ownership is maintained in the links between the Asset_Owner and Owner tables, but the Bill_table only lists a single owner, the primary owner who receives the bill. The presumption is that that owner is responsible for seeing that the taxes are paid; if you send the same bill to all owners, you might receive multiple tax payments that would have to be refunded.

The second key relationship is between owners and assets. This relationship is many to many because a single owner may have many assets and an asset may have more than one owner. Relational database design requires a composite, or link, table in this situation, and that table is called Assets_Owner. Each asset is joined to one or more entries in the Assets_Owner table. If an asset has more than one owner, there will be multiple records in the Assets_ Owner table to represent that. For these records, the Asset_Id will be the same, but the Owner_Id will be different; there will be one record in the

Assest_Owner_Table for every owner of this particular asset. This is how you get around the issue of how many owners to assign to an asset; it may have as many owners as it needs, but it must have at least one owner. An asset must have an owner, even if it is a dummy value that has an Owner_Id of 99999 that is linked to a dummy owner whose name is Unknown with an address of Unknown, and so on. Asset_ID and Owner_ID are foreign keys in this table; the primary key for Assets_Owners table, Asset_Owner_Id, is not shown but would exist. Tables whose primary key is not a foreign key in another table technically do not need to have a primary key, but it is good design to have one.

Another relationship, Owners table to Asset_Owner table, is a zero/one-to-many relationship. This means an owner may have no assets; these are the owners we wish to keep in the system even though they currently may own no assets. The relationship from the Assets_Owner table to the Owners table is a mandatory one-to-one relationship. No asset/owner combination may have zero owners but may have multiple owners.

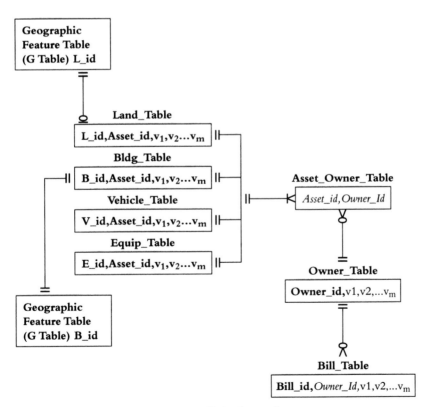

Figure 3.3 Detailed schema diagram.

The relationships between bills and the assets themselves (land, vehicle, buildings, and equipment) are a little more complex. Clearly, owners with no assets will receive no bill, and owners with many assets will receive only one bill. So the relationships between the bill class and the asset classes are one to many and nonmandatory, but at least one of the relationships must return a value. Individually, the relationship is one bill to many assets (optional), but collectively it is one bill to at least one asset. This possibility shows that a composite table is clearly needed between owners and assets because owners may own many assets and assets may have more than one owner.

The geographic feature tables exist only for the land parcels and the building outlines. Vehicles and equipment get moved frequently, so their location is not a permanent property and is not modeled in this database. The land parcel geographic table has a zero-to-one relationship with the land parcel data table. A zero relationship means that a tax bill will not be generated for the land asset, which would be the case for properties owned by the various levels of government and some other organizations. Those land parcels will exist in the geographic table, however. The Land_Table-to-Geographic Feature Table relationship is one-to-one and mandatory. All records in the land data table are linked to one, and only one, record in the land geography table. This may not be a realistic situation; a parcel of land that is considered a single feature for tax purposes may consist of two or more separate polygon features. In that case the relationship would be one-to-many and mandatory. Every record in the land data table has a corresponding record in the land geography table. This means we know where all the taxable land is and have it digitally mapped. The relationship between the building geography and the data is simpler: one to one mandatory in both directions. Also notice that there is no link between the two geography tables, although there could be. The Land Geographic Feature Table-to_Building Geographic Feature Table relationship is a zero, one, or many relationship because a land parcel may have no building on it. The Building Geographic Feature Table-to-Land Geographic Feature Table relationship going the other way would be a mandatory one-to-many relationship because building lines can cross property lines, creating a situation where the building is associated with more than one land parcel.

Step 4. Broadly sketch out the key pieces of information you need to know about the members of each class of objects. Fields are always easy to add to tables, and spending a lot of time at the beginning of the process in detailing exactly how and what you are going to record in each field of a table is not necessary. It is important, though, to have a general idea of what kinds of information are important for each table. As a general principle, include only the information that is specific to the type of object. This is why the geographic feature tables shown for the land and building object classes are separate. They contain only the geographic information (location, area); all the other information about the features is maintained in the asset tables.

Step 5. Sketch out the detailed schema diagram. This is best done on a piece of paper where you can move tables around and erase mistakes easily. This example (Figure 3.3) uses a notation called information engineering symbols

(IES). There are other notation systems for database design (e.g., Universal Modelling Language); the database world has not settled on a single format yet. The formal design can also be accomplished using CASE tools, which sometimes come with GIS software packages. Although a complete schema will eventually include every field for each table, this can take up lots of room on a piece of paper. The example shown includes only the primary keys (show in bold) and foreign keys (shown in italic). Of course you will know other pieces of information about each feature in the tables; these are shown as v_1, v_2, and so on. The design concerns for this attribute information are covered in the next section.

A data dictionary that describes the contents of the various tables in the database and a schema are the central design elements at this stage. Tables and their resulting dictionaries can get quite large, and schemas can require large-format printer or plotters to display them. They are essential documents and contain all the information a database professional needs to understand the structure of your database. A schema can take months to complete and be much more involved than the simple example we presented here. As you go through the process, you need to keep several key issues in mind as you design the schema:

- Tables should be specific to the objects they represent and contain information only about that class of object. So a table of utility poles should not contain information about the objects that might be on the poles, (e.g., transformers). A land parcel table should not contain information about the owner; all it needs is a foreign key that can link the land parcel table to the owner table. The land parcel is a geographic feature, and that is the information you store in that table. The owner or owners are people or entities and have different properties, so they belong in different tables.

- Often a relationship is one to one, but in a few instances can be one to many. Even if the situation occurs only once in the database, it must be explicitly modeled in the schema.

- Many-to-many relationships require a composite, or link, table. These tables contain only a primary key for the table (which was missing in the example we presented) and foreign keys to the tables containing the features involved in the many-to-many relationship. Composite tables may never appear in screens that users see, but they must be in the database to maintain this most complex of relationships.

If you are implementing your GIS using a commercial RDBMS, the various types of relationships (one to one, one to many, etc.) may be referred to with different terminology. The software will probably also have tools for creating a database schema. Ultimately the schema is implemented in the database through SQL. Each table is set up with SQL statements, and the relationships are defined with SQL. The CASE tools allow you to visually draft the schema that best represents your view of the world and then automatically generates the SQL statements that will create the database and that become a textual representation of the schema. Those are usually quite technical steps, but the initial work of identifying the tables and the relationships between them can be done in a group setting with nothing more than a large piece of paper and a marker. A database professional should be able to take that and generate the necessary SQL to create the database. The visual representation of the database, as we said earlier, is a blueprint and, like blueprints, is constantly consulted during the actual construction process.

Metadata

Metadata is usually described as data about the data. It is the information you need to document your data set sufficiently so that an outsider can understand all the key issues involved in the construction of the data set, what the various values in the data set mean, what projection you are using, and so on. One analogy is to a catalogue card for a book in a library, although most metadata is considerably more involved than that. The product of the process of creating metadata is a file that describes your data set, or pieces of your database. The mountains of information available on how to produce it, what to include, how to check it against a standard, and publish it is huge, but the actual product is rather small.

The likelihood that this chapter will be out of date by the time it hits print is high because there is a lot of work being done around the world to implement standards and also because the techniques to disseminate metadata are expanding rapidly. Early implementations of geographic metadata were simple text files that could be read by word processors. The National Spatial Data Infrastructure (NSDI) initiative of the U.S. government led to a standard for implementing metadata using hypertext markup language (HTML). Currently, the push is to using extensions to HTML for producing metadata, Simple Graphics Markup Language (SGML) and Extensible Markup Language (XML). It is almost certain that there will be new high-level languages developed out of these that will have additional advantages for implementing metadata.

Since the 1990s there has been a proliferation of metadata standards, but in recent years national organizations have begun to cooperate on a set of international standards that may eventually make it easier to document and share geographic information across the world. The United States Federal Geographic Data Committee (FGDC) standards were an early version of metadata standards, and now there are groups in Europe, Australia and New Zealand (anzlic.org.au/asdi/metaelem.htm), and internationally that are working on geographic metadata. Additionally, the Dublin Core is a project that is attempting to integrate metadata efforts across disciplines and digital data types so that groups working on metadata standards for image data, for example, will have some relationship to standards developed for other types of data.

The FGDC has been at the task the longest of any of the organizations (see *How They Did It — U.S. Federal Geospatial Metadata*), and the standard is mature and well disseminated. It consists of seven information segments, three supporting sections, and major content areas (Table 3.3). Within each content area are often dozens of elements that specify the details of the content. These details are impressive; there are more than 300 elements, including 199 data entry elements. However, the standard has both required and optional elements and only certain items, fewer than 20, in the Identification and Metadata Reference sections are mandatory. Creators must fill in these sections, and some GIS software systems will read through the data set and produce a template for many of the optional items as well. For example, if the map projection for the data set is available in a file formatted for the software to read, some software will read that information and place it in the appropriate metadata location.

How They Did It – U.S. Federal Content Standards

How the current U.S. federal metadata standards came about is an example of the slow, but careful, process of governmental coordination and cooperation. In the late 1980s the United States Geological Survey (USGS) set some requirements for how descriptive information about digital geospatial data should be collected and even before the issuing of Office of Policy and Management (OPM) Circular A-16 in October 1990 was circulating and discussing draft metadata documentation standards in the GIS community. The OPM circular established the FGDC with 12 federal agencies on the coordinating committee and the Department of the Interior as the lead agency. This new committee (FGDC) began work in 1990 in the Department of the Interior's USGS. This unit of Interior was the logical location because of the widespread use of their digital data at several different scales by many other federal, state, and local agencies. Executive Order No. 12906 of the U.S. federal government in April 11, 1994, is the central document, and it took 4 years of coordination, negotiation, and work by the FGDC before it could be issued. It required federal agencies and organizations receiving federal funds to document their geographic information using the FGDC's Content Standard for Digital Geospatial Metadata. There were three goals for this order:

- Minimize the costs of creating geographic information. By forcing the creation and distribution of metadata, different agencies of the government are more easily able to search the resources of other agencies before they expend time and money creating a near duplicate of a data set that already exists.

- Encourage cooperative collection activities. The existence of metadata for a set of geographic data actually assists in the creation of the data (i.e., you have to be able to document the data, which forces you to consider aspects of data design you might otherwise avoid).

continued

- Establish a national framework for geographic data. This order also established the National Spatial Data Infrastructure to design a system for creating metadata-based sites on the World Wide Web for users to seek out spatial data in an organized way.

The deadlines established in the order clearly show that most of the work had been already done before the order was issued. The committee held a public forum as early as 1992 and circulated draft standards in 1993. The first standards were published in June 1994, only a few months after the executive order was issued. After that publication a considerable amount of the resources went into educating the community about the standards and establishing partnerships with state, local, tribal, and university data developers to document their data with the new standard and to set up searchable sites under the NSDI for sharing the data. These partnerships were funded with grants from the U.S. Department of the Interior, through the FDGC, and have been spread all around the country. Staff members of the FGDC were very active, going to GIS conferences around the country and giving presentations on the content standard and how its use was going to make geographic data more usable and accessible.

Adoption of the content standards, except for federal agencies who must use them, has been slow, as has the growth of the NSDI; as of May 2002 there were only 242 nodes on the NSDI, which, given the size of the GIS community, is rather small. Ten percent or more of the servers in the NSDI are likely to be down at any time, and establishing an NSDI site is not something a casual user will do. At the time of writing the position of metadata coordinator on the FGDC was vacant. The Bureau of Land Management was an early adopter and diffuser of metadata and tools to create it, but the Web site that deals with the NSDI had not been modified since September 1998.

Table 3.3 Metadata – Federal Geographic Data Committee

Major Content Areas – Federal (U.S.) Geographic Data Committee Standards

1. Identification information	An abstract of detailed information in the other sections.
2. Data quality information	Assessment of the accuracy of the spatial and attribute data being described.
3. Spatial data organization information	Detailed documentation of the types of spatial features in the data set. The feature types, vector and raster, must correspond to the Spatial Data Transfer Standard, which was developed along with the metadata standards.
4. Spatial reference information	Coordinate system, projection, and geographic extent information, including information about elevations or depths.

Major Content Areas – Federal (U.S.) Geographic Data Committee Standards

5. Entity and attribute information	Information about the attributes attached to the features; a detailed data dictionary explaining the values in the data fields.
6. Distribution information	How to obtain the data.
7. Metadata reference information	Description of the metadata itself and how it was produced
8. Citation information	How this information should be referenced if others use it.
9. Time period information	Period of time over which the information was prepared and whether it is updated or not.
10. Contact information	How to reach the custodians of the data.

Required Elements

1. Identification Information	*Originator*: name of an organization or individual that developed the data set.
	Publication_Date: the date when the data set is published or otherwise made available for release.
	Abstract: narrative summary of the data set.
	Purpose: summary of the intentions for which the data set was developed.
	Calendar_Date: the year (and optionally the month or month and day) for which the data set corresponds to the ground.
	Currentness_Reference: the basis on which the time period of content information is determined.
	Status: the state of the data set.
	Maintenance_and_Update_Frequency: the frequency with which changes and additions are made to the data set after the initial data set is completed.
	Theme_Keyword_Thesaurus: reference to a formally registered thesaurus or similarly authoritative source of theme keywords.
	Theme_Keyword: common-use word or phrase used to describe the subject of the data set.

continued

Table 3.3 (*Continued*)	
Required Elements	
	Access_Constraints: restrictions and legal prerequisites for accessing the data set.
	Use_Constraints: restrictions and legal constraints for using the data set after access is granted.
10. Metadata reference information	*Contact_Organization:* organization responsible for the metadata information.
	Contact_Address: four required elements of the mailing or physical address of the contact organization.
	Contact_Voice_Telephone: telephone number of the contact organization.

Minimal required metadata documentation is not the time-consuming task that many data creators think it is. Although these metadata efforts are all tied in somehow with development of the International Organization for Standardization (ISO) standards, it is not as clear how well they are tied to the activities of the Dublin Core. This is partly due to the different backgrounds of the participants; the GIS metadata standards have been developed largely by bureaucrats of national governments working either alone or together, but these professionals have principally come from the geographic data community, that is, they are users and producers of geographic data. The impetus behind the overarching efforts of the Dublin Core comes from the international librarian community, and it is being staffed and organized by professionals whose concerns are the organization, storage, and retrieval of information.

The professionals behind the Dublin Core have expressed concern that the many groups, not just producers and consumers of geographic data, working on metadata standards are going to get so widely separated from each other that they will have little in common. With those concerns, they have developed a 15-element metadata standard that is designed to accommodate many different forms of digital data, not just geographic data (Table 3.4). Although the FGDC standard has a hierarchical structure with many sub-elements and sub-sub-elements under each major section and has some strong restrictions on how data are presented, the Dublin Core is a much simpler standard. There is no complex hierarchy of elements with long names but only these 15 identifiers with suggestions to use standard lists of elements such as standard lists of data types. Examples are Multipurpose Internet Mail Extensions (MIMEs), describing the data type; Uniform Resource Locators (URLs), documenting the Internet location; and International Standard Book Number (ISBN) as a unique identifier. The Dublin Core is a much easier standard to implement and has the added advantage of being usable for all sorts of other data (e.g., documents, nongeographic databases, and

Table 3.4 Fifteen Elements of the Dublin Core (7/2/1999)

Title	Name of the Resource
Creator	Entity (person, organization or service) creating the resource.
Subject	Subject and keywords; they suggest using formal keyword systems.
Description	Abstract, table of contents, free-text description of the resource.
Publisher	Person, organization, or service responsible for publication,
Contributor	Entity (person, organization, or service) contributing to the resource.
Date	Typically, the date of creation or availability of the resource.
Type	The nature or genre of the resource; suggested use of the Dublin Core Type Vocabulary Collection, Dataset, Event, Image, Interactive Resource, Service, Software, Sound, Text. Geographic data would be a data set.
Format	Digital encoding of the resource. The suggested list (MIME) does not yet include any geographic data formats.
Identifier	A unique identifier such as a URL or ISBN.
Source	Reference source for the resource (i.e., where it came from).
Language	Language of the resource content.
Relation	Relation to another resource.
Coverage	Suggest use of named places and time periods rather than sets of coordinates or dates.
Rights	Information about who holds the rights to the resource.

Source: dublincore.org/documents/1999/07/02/dces/ This is a DCMI Recommendation, copyright © 2002 (Dublin Core Metadata Initiative).

images). The committees behind the Dublin Core recognize that geographic data are, after all, only digital data and can be documented using a simpler standard. GIS practitioners will recognize, however, that it fails to include necessary documentation for geographic data such as map projection information and other database documentation normally found in data dictionaries. Map projection information is easy to include in the coverage section, though, and if the nongeographic or attribute data are organized in a relational database management system, it is possible to include the data dictionary in tables in the database itself. The Dublin Core itself is probably not adequate for documenting geographic data, but with a few additions it provides a simpler, more generalized format for creating metadata.

In addition to these national and international standards in varying stages of development, many organizations (mostly governmental) have created their own formats for metadata, and some GIS software vendors have also produced formats for documenting data. In fact, it is the proliferation of these varying standards that led to the national and international attempts to standardize the standards. As the more widespread standards emerged, people began to develop tools to help users implement the standards.

In the mid- to late 1990s there was proliferation of these tools. In the United States, FGDC was pushing users and vendors to document their data, and individuals in federal agencies developed and distributed tools for users to create and validate FGDC metadata. These were sometimes software-specific but often were stand-alone programs to assist in the creation of metadata. Most of these tools were labors of love on the part of the creators and have not been maintained or updated, although there are many who are still using such tools as the metadata Arc Macro Language (AML) script written by Sol Katz in the Bureau of Land Management.

Now metadata creation tools are more likely to be built into the software used to create the spatial data. There is also a move toward the direct incorporation of metadata into the database. Previously you had to locate the metadata somewhere else and place the file somewhere in close proximity to the actual data. The reality is, though, that many users are confused about metadata, resistant to spending the time to develop it, and not certain of its utility.

The adoption of standards for metadata has been slow, and many practitioners still do not document their data or do it very poorly. It seems as though the rate of development of the standards is far exceeding the adoption of any standard. Users or organizations unwilling to document their data to a standard can use this rapidity of change as a reasonable excuse. This is unfortunate because metadata answers so many questions about data you have obtained from others and will answer their questions if you provide it to them. The problem is that metadata, prepared to almost any standard, is difficult to create, and it takes specialists to do it correctly. It has been likened to cataloguing in libraries; it requires a specialized set of tasks and not every librarian is very good at it. But organizations with insufficient staff to set aside all or part of one professional to become the metadata expert in the organization face an uphill struggle. And consulting firms who come into an organization, develop some data, and leave will have no incentive at all to spend the time to create metadata unless the requirement and resources have been explicitly included in the contract, and they often are not. The documentation of all geographic data sets to a metadata standard is a generic goal of the world of GIS practitioners, but practically it lags behind and is the last thing done.

A complete database schema will contain much more information than merely a data dictionary and table/relationships diagrams. The elements listed in Table 3.1, such as queries, reports and forms, optional metadata, and so on, are all part of a schema, but they are generated from the data dictionary, the tables, and relationships you have defined. Those items are at the core of your schema, and a well-designed database can support a very wide range of reports, input forms, queries, and workflows. It is the design of the tables and the relationships between them that is at the core of the schema, and that is why we have focused on them here.

GIS implementations almost always involve existing databases and their schemas. The decision of whether to try to bundle all the different databases together, which is something GIS is particularly well suited to, or to redesign the entire system into a single database is important. GIS implementation is always a good time to look at the structure of all your databases, and sometimes the right decision is to redesign the system completely and move the existing data into the new schema. But whether you choose to take that approach or take the approach that links the databases together, you will need to understand the schemas of the existing database. At an early point in the process someone will always say, "We don't have to reinvent the wheel here," that is, we have this perfectly acceptable database that currently meets our needs; let's just tie in a GIS. However, sometimes the wheel you have is not particularly round and runs somewhat awkwardly. In those situations it is a good idea to sit in a room for a while with a large blank piece of paper and sketch out a database that might really work for you. Some wheels are better than others, but you have to design them carefully, and that starts with understanding your existing schema or creating a new and better one.

ADDITIONAL READING

Chandler, A., D. Foley, and A. M. Hafez. 1999. Federal Geographic Data Committee (FGDC) Metadata into MARC21 and Dublin Core: Towards an Alternative to the FGDC Clearinghouse. Lafayette, LA: University of Louisiana. eeirc.nwrc.gov/pubs/crosswalk/fgdc-marc-dc.htm.

Federal Geographic Data Committee. 1998. Content Standard for Digital Geospatial Metadata. Washington, D.C.: United States Government Printing Office.

Fraser, B., and M. Gluck. 1999. "Special Section — Usability of Geospatial Metadata or Space-Time Matters." *Bulletin of the American Society for Information Science* 25 (6):24–32.

Harrington, J. L. 1998. *Relational Database Design Clearly Explained.* AP Professional: San Diego, CA.

Konkel, G. 1999. *Final Completion Report: Snohomish Basin Literature Review and GIS Data Acquisition Project.* Washington Department of Transportation Environmental Affairs Office: Olympia, WA. wsdot.wa.gov/eesc/environmental/programs/watershed/snobas/other_links/final_report.cfm.

Smits, J. 1999. "Digital Cartographic Materials." *Cataloguing and Classification Quarterly* 27(3): 321–343.

INTERNET RESOURCES

Metadata Publications, FGDC:
fgdc.gov/publications/documents/metadata/metadata.html

Department of Interior, Bureau of Land Management's Metadata and WWW Mapping Homepage:
blm.gov/gis/nsdi.html

International Organization for Standardization (ISO) TC/211 —
Geographic Information/Geomatics:
isotc211.org/

GIS and Metadata: Frequently Asked Questions:
standardsinaction.org/gismetadata/FAQMetadata.htm

Designing Spatial Data

T he real world is a complex place. Hardly an original thought, but one that you
need to keep in mind as you design your GIS. As an example, consider the reality
of a land parcel. The parcel is out there in the real world; it exists. There may be people who live in buildings on the parcel or conduct a business on the parcel, or it may
just lie unused at the moment, slowly returning to its natural state. The parcel is a real
thing, but there are reasons to model this land parcel in different ways because interacting directly with the parcel is often difficult. It is easier to deal with one of many
models of the parcel.

Choosing the Appropriate
Mix of Data Models

One model of the parcel is its legal model, the deed. That deed records the geographic descriptors of the parcel, a sketch of where it is located relative to adjacent
parcels' ownership information (see Figure 4.1). This model of the parcel is printed
on an 8.5- by 11-inch or legal-size piece of paper, and copies of it exist in many
places. The purpose of the deed model of the parcel is to facilitate transfer of ownership. The information we need to extract from the parcel and its owners is all related
to the legal aspects of the buying and selling of the parcel.

Another model of that land parcel is in the community's assessor's database. The
purpose of this model of the parcel is to assess property taxes due in a given period of
time to the taxing governmental unit. This model often has a digital (computer) as
well as analogue (paper) form. The assessed value of the land, improvements that have
been made to it, and the address of the owner(s) are all pieces of information that the
assessor has abstracted from the parcel and entered into a database. From that database
the assessor generates tax bills, the grand list for the public, and other standard reports
necessary to the functioning of the government. Without a standard model of a land
parcel the procedure for assessing land would be ad hoc, capricious, and illegal. The
design of the model and the systems that use it provide taxing governments with a
good way to demonstrate that the assessment is fair. Without a clear model of the land
parcel, it would be impossible to do this.

Piece or parcel of land with the buildings and improvements thereon, situation in the Town of West Hartford, County of Hartford and State of Connecticut, known as No. 16 White Ave. and also being shown and designated at Lot. No. 24 on a certain map or plan entitled, "Revised Map of Property of Carl G. Dahl West Hartford, Conn:," which map or plan is on file in the Town Clerk's Office in said Town of West Hartford, reference to which is hereby made, and being bounded:

NORTH by land undesignated on said map or plan (being land now or formerly of Wilbrad Desroches, et al, and land now or formerly of Ivan H. Skoglun, in part by each, in all), 50 feet;

EAST by Lot No. 25, as shown on said map or plan, 118.50 feet

SOUTH by White Ave., 50 feet; and

WEST by Lot No. 23, as shown on said map or plan, 119.06 feet

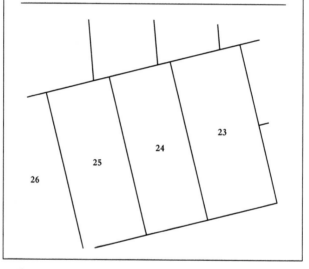

Figure 4.1 Legal and GIS representation of a property.

The land parcel can also be modeled as a feature in a layer of a municipal GIS. Although the purposes of the legal and assessor's models are quite specific, the model of the parcel in a GIS often has to serve the purposes of several different users. Frequently the GIS model of the parcel is used for mapping and property comparison by the assessor, but many other units of government — planners, engineers, public safety and safety officials, for instance — may need to know the location of the property as well. The designers of the assessor's database had clear purposes in mind during the design process, and these purposes are reflected in the database's structure as well as the planned output reports and input forms. The model of the parcel in a GIS must similarly be designed to meet the needs of the users.

Hierarchial Level	Representation	Discussion
Real World		Identification of which real-world features to model is an important early step because it is not possible to include all elements. A black and white aerial photograph example is already an abstraction (color removed, vertical point of view) of the real world.
Data Model	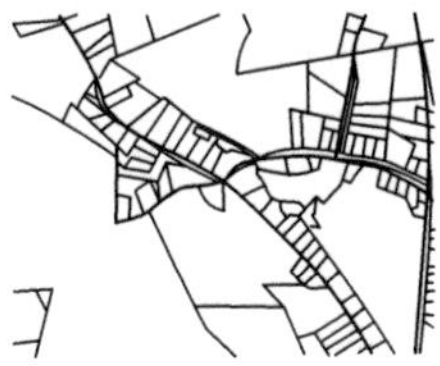	Selection of mix of data models is driven by user needs. Topologic vector data model and raster data model are common choices, although there are other specialized data models. The data model can usually be represented as a sketch or picture.
Data Structure	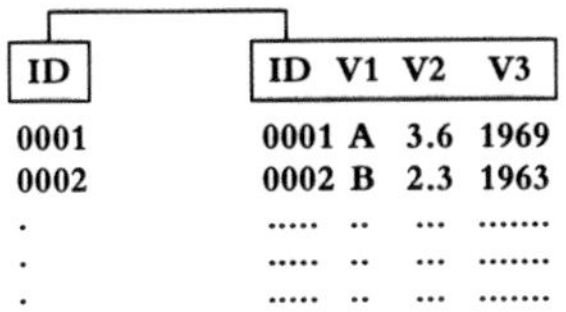 ID ID V1 V2 V3 0001 0001 A 3.6 1969 0002 0002 B 2.3 1963	A tabular representation of the data model. The data structure is a schematic representation of how the data will be stored for retrieval by the GIS.
File Structure	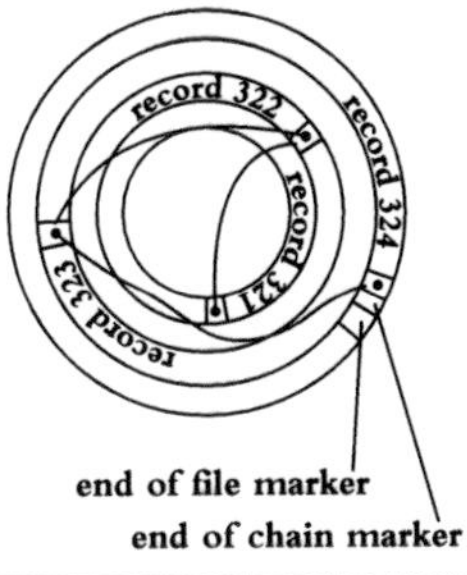 **end of file marker** **end of chain marker**	The physical storage of the data structure in the system. The file structure is often proprietary and unique to the particular GIS, but some are beginning to use the file structures of commercial relational database management systems. An appropriate file structure is essential for rapid retrieval of geographic data.

Figure 4.2 GIS data abstraction. Adapted from Peuquet in Peuquet and Marble 1990, Figure 2, p. 254. Used with permission.

The hierarchical process for abstracting items in the real world and getting them into the computer has been very well known for some time (Peuquet in Peuquet and Marble 1990) and came out of the development of relational databases in the 1970s (Figure 4.2). As you move down the hierarchy, components become more structured and simplified. Reality is difficult to discuss, and people use a mixture of words, pictures, and gestures to describe a feature in the real world. When you have settled on a model of the real feature, it is often possible to describe that feature in a diagram or map. The two most common data models in GIS are the vector model, in which you diagram the feature as a point, line, or polygon, and the raster data model, in which reality is abstracted as a set of pixels or rectangular units. At the next level, the data structure, you must specify the details of that diagram or map. If you are choosing to model the land parcel as an assessor's record, each element of that record must be described based on the kind of information will be in it, the maximum/minimum number of characters or numbers that will be allowed there, and so on. The data structure of the land parcel at this level is a detailed written description that usually follows a standard set of rules and a set of tabular data locating the parcel in a coordinate system. A feature like a swamp is described in a raster data model by stating the height and width of each cell in some coordinate system and what numeric or text value indicates that the cell represents part of a swamp. Finally, this information needs to be stored digitally in the computer so it can be retrieved for use. Are you going to store the data as a string of numbers or will it be stored in some kind of dimensional array? Are you going to store it in a file format that many software programs can read directly or will you store in some proprietary format that only selected programs and users can read? At each level of this process, reality, data model, data structure, and file structure, there are important questions to ask and answer in the design process.

Choosing a Subset of Reality

No GIS or database can include all aspects of reality; you always have the task of choosing a subset of real-world features that you will model in your GIS. For example, an electric utility has a view of the world that is quite different from a visiting nurse association, and their different missions and goals require that different elements of the real world be modeled in their geographic information systems (see Table 4.1).

Knowing what you need to model in your GIS and, as important, what you do not need to model is the critical first step in the design process. The group process of restricting your view of reality so that you can focus on what is really important to meet the organization's goal is very important. Failure to restrict your view of the world sufficiently means you will be including a lot of expensive information you do not need. Too much restriction of your view of the world means you will leave out information critical to your mission. For example, is it necessary to know the exact location and information about each utility pole, or is it enough to know the average number of poles per mile of distribution line? If your only purpose is to estimate the number of poles in a community for inventory tax purposes, the latter may be sufficient, but if you want to manage the poles for a replacement program, you will need to know details about each one, including its location. In the visiting nurse association example, only the topological relationships (discussed later in this chapter) in the road network are necessary to reach a client: take Brown Street to the intersection of Booth Street and turn right. But an electric utility needs to capture the road network where it really is because their distribution network closely parallels it. Both organizations need a road network with topological relationships, but how they model the reality of the road network will differ. Both will need it for routing practitioners to clients or maintenance

Table 4.1 Comparing Realities

Generic Classes of Real-world Features	Visiting Nurse Association	Electric Utility
People	Clients with medical needs	Customers with electrical power needs
Things	Portable medical equipment that must get to the people who need it	Fixed and portable equipment for generation and distribution of electricity
Geographic features		
Areas	Service areas of the organization and other related service providers; governmental units in service area	Service area of organization and competing providers, usually set by regulation; maintenance regions; governmental units in service area; substations
Lines	Road network for access to clients and services	Distribution lines: generation to substation, substation to clients; road network for access to clients and facilities
Points	Clients' residences, practitioners' offices, hospitals, nursing homes, rehabilitation faculties, drug stores, medical supply providers	Service points, substations (for modeling distribution), utility poles

crews to problem sites, but the utility will also need a more detailed and accurate representation, including the road's rights of way, for installation of poles. The visiting nurse association will only need a representation of the road as a linear feature connected to other linear features, and there is no need for high levels of spatial accuracy.

The Two Principal Data Models

Practitioners of GIS have for years grouped geographic data into two classes of models: raster and vector data models. The choice of raster or vector data models for a GIS has usually been presented as an either-or question and sometimes as a debate over the merits of the two data models (Goodchild and Kemp 1990). This view has presumed that there is one data model that is better than others and that the data models are like prizefighters: each exploits the weaknesses of the other and uses its strengths to represent the world in a better way. More recently, however, practitioners have realized that their needs may incorporate multiple models for spatial data, and it has become unnecessary to view the process as an either-or choice. Rather, you need to consider your needs for modeling your portion of the real world. Many, but not all, GIS software systems can move data back and forth between these two data models, and you may discover that certain tasks are easier to perform in one or the other of the data models. But many users find that one or the other of the data models meets their needs so that one is best for their applications. Organizations that require high-quality mapped output from their GIS will find that only the vector data model allows that. Users who will be doing lots of overlaying of different geographic features will find out that the raster data model is more efficient for that type of task.

When you are trying to develop a model of a real-world object, it is vital to keep several questions in mind. By knowing the answers to these questions you can ensure that your model of the real-world objects will relate successfully to the object.

What Is the Purpose of This Model?

If the purpose of the model is solely to be a digital representation of the feature for mapping, the design issues are rather simple. You will need to decide what kind of digital feature you will use to represent the real-world feature and how to symbolize it on the map. In the case of a land parcel, if the sole purpose is mapping of the land parcel for visual representation, the parcel will consist of a set of lines with symbolization and a point inside these lines where some unique identification number is attached. There may be one symbol representing the coincident features of a road right of way and a parcel edge and a second line symbol for the other edges of a property. The point symbol may not be visible at all because its only purpose is to serve as an anchor for some text. The digital line, point, and text features may exist separately in your GIS because the map reader does the linking between them visually. The presence of a number inside a set of bounding lines informs the reader that that number identifies the polygon represented by the bounding lines. This is a common way to visually represent land parcels using CAD tools.

If the model is to serve multiple purposes, design concerns become more complex. Using the land parcel example, let us say that in addition to simple mapping you want to be able to link the assessor's information so that this information can be included in a report that provides a quick printout of the important information along with a map of the parcel and adjoining parcels. In this case, the land parcel can no long exist only as a set of lines but must be modeled as a polygon or area in your GIS. In addition, the unique identifier (parcel identification number, or PIN) must be in the GIS database and directly linked to the parcel so that it can be connected with the assessor's information. The system, not having the ability to see the PIN inside the lines bounding the parcel, must know the parcel's PIN and relate it to that parcel and only that parcel.

The representation or model of a real-world feature may be different depending on the purpose of the model. Although a land parcel in the real world is a polygon, there may be some value in representing that polygon as a point in some databases. The fire department may need to have the buildings on land parcel represented as a digital image so that they can see the buildings and where they are relative to each other when planning or evaluating a fire response. But when they want to know what hazardous materials are stored in that building and where in the building they are stored, the building may need only to be a record in a database linked to an address or a point. A business responsible for distributing goods over a wide area may be able to find good locations for a warehouse by modeling their regional demand and supply facilities as points along a network of major highways, but when they have to actually move goods from a particular warehouse to a set of locations, they will need a larger-scale model of the transportation network. So the same real-world feature may exist in your GIS database several times for the different purposes. The need to display or analyze geographic data at very different spatial scales also results in this apparent duplication. A state agency trying to represent all the land parcels in the state may be forced to model each parcel as a point to even begin to display and locate the hundreds of thousands or millions of parcels that exist. At the state scale that may be the right choice, but at the local level modeling the parcels, as point will not suffice. The parcels will have to be modeled as polygons. At one scale you might want to represent only major arterial roads in your database, but at a larger scale you might want to include all roads.

Which Data Model, Raster or Vector, Better Suits My Purpose, or Is a Combination Appropriate?

Traditionally the highest-level model choice in GIS has been presented as the choice between the raster (grid) and the vector model. A diagram such as Figure 4.3 is standard fare in any introductory text on GIS and is certainly not out of place here. The raster model represents the real world as a *tesselation*, with grid cells having a certain length and width and covering the entire rectangular area of interest. The value in the cell represents the real-world feature at that location within the cell. The real world is not modeled as set of points, lines, or polygons but as a complete covering of regular (usually square) cells. A point can be located only within a cell, a line is a collection of connected cells underneath the feature, and a polygon or area feature is a set of connected cells touching the polygon.

The vector data model represents real-world features as strings of x/y pairs representing spatial information about the features. A point is a single pair of values with an x value representing the longitude or x dimension of the point in some coordinate space and a y value representing the latitude or y dimension of the point. Linear features may consist of a set of connected points or some mathematical function describing the beginning point of the feature and the formula that constructs it. All GISs that support the vector data model can model circles and other curved lines as sets of short, connected straight lines. Some are able to store information in such a way as to represent the feature as a curve. If that is important in your model of the world, you need to ensure that the software you select can treat features that way. Polygon or area vectors model the real-world feature as a connected set of lines that closes on itself. These features are discussed in more detail later in this chapter. For all of these three basic feature types in the vector

	VECTOR	RASTER	CONSIDERATIONS
Point Feature	•	▦	Indeterminate location within pixel (raster)
Line Feature			Aliasing or stair steps, length calculation (raster); true arc or approximations with many straight lines (vector)
Area Feature			Raster cell size (resolution)

Figure 4.3 Raster-vector comparison.

data model you can imagine that a real-world feature might consist of multiple instances of points, lines, or polygons. In a pavement management system a section of pavement may consist of several segments of street that together make up the management section. If the purpose were only mapping, you could represent each segment between intersections as single lines, but for managing pavement you would need to represent those multiple lines as a single thing. In a similar fashion, governments frequently have noncontiguous pieces of territory such as islands off a coastline, and you can combine these into a single multipolygon region.

Most GISs are able to handle both raster/image data and vector information; so the correct question is what is the proper data model for particular features rather than what is the proper data model for the entire GIS. This discussion ignores the quad tree data model, which is the third principal model for geographic data; some consider this a submodel of the raster data model. This data model is still a topic of research and there are a few commercial software systems that use it for geographic data; advances in computer processors and storage have removed a lot of its early appeal.

There are five possibilities for combining the raster and vector data models that you need to consider when implementing a GIS:

- *Principally vector with raster underlay.* This model combination uses the vector model as the principal data model, but for context there is a raster backdrop. This backdrop is often a digital version of aerial photography, a digital orthophoto, but sometimes is processed satellite information. This option is becoming more common as compression techniques for the background digital images improve, allowing rapid redraw and panning. This option assumes that you will be acquiring the raster backdrops in some standard format and do not plan to do any processing or changing of that backdrop. This topic is discussed in greater detail in chapter 6.

◆ *Principally raster with vector overlay.* This model combination is useful when remotely sensed imagery is the primary source of data for your GIS or you need to directly process any raster information. Software systems marketed principally as remote sensing applications all have the capability of draping vector information on top of the rasters, and the distinction between GIS and remote sensing software systems of this sort is becoming more blurred. The vectors provide a map-like context for the raster data. Because raster data sets are rectangular, having vector polygons to draw over the raster data allows you to put roads, governmental boundaries, water features, and other base map information to provide locational context for viewers. If this is your combination choice, it implies that you have no need of vector modeling capabilities (i.e., topological data sets). It assumes that all the processing and modeling of reality is taking place in the raster data model, and vector information is only symbolized and drawn over the raster pixels in the correct location. Using vector polygons to clip out segments of the raster, for example, would require topology in the vector data set and would mean you would need software to fit the fifth combination of data models.

◆ *Solely raster.* This option, although logically a possible one, really does not exist because all the raster GISs and remote sensing software systems are capable of draping vector data over the grid cells.

◆ *Solely topologic vector.* In this data model combination all the data are stored as vectors, and topological relationships of connectivity and adjacency are known or can be computed. This is the most map-like of the combinations in its output but, like the solely raster option, really does not exist anymore because the capability to display raster images behind vector data is so widespread and useful.

◆ *Full vector and raster.* If your organization will be processing its own raster data, whether for backdrop or modeling purposes, and also needs topologic vector information, this combination of data models is necessary. It can deal with both data models, and also the data to be moved between the data models. In some cases it is easier to analyze the real world in the raster data model and transform the results of the analysis to the vector data model for further analysis or display. More than 10 years ago a software vendor claimed that in the near future GISs would have these capabilities, and the system would decide which was the appropriate data model to use, transform the data into that model, and do the analysis or query. This has not yet come true, and it is still up to the users of systems with these dual capabilities to decide what is appropriate in what circumstances.

So the raster/vector decision is not a debate with a winner. Nor is it appropriate to state, as has been common in the literature, that the raster data model is a better fit for natural resource applications and the vector data model for artificial information. Rather, certain features of reality are easier and better to handle in one or the other models. Generally, if you have any needs for aerial photography backgrounds or satellite image data in your GIS, you will need some raster display capabilities within your vector GIS. If your needs for raster information are greater than this (e.g., you need to acquire and process remotely sensed satellite data), you will need software and hardware specific to those tasks. The remainder of this book will focus on the process of design and implementation for the topologic vector data model with raster backdrop needs.

Layers and Objects

The debate between layers and objects has been ongoing in GIS for some time, at least a decade (Goodchild and Kemp 1990, Unit 22). This, however, is a true debate and deserves discussion. Early GISs all used the layer as the principle data structure, and this came from using layers to produce maps. Manual cartographers tend to work in layers even when the final map is going to be entirely in black and white. There may be one layer for each width of line work or a layer with text information and a layer with fill patterns. Drafted on clear acetate, each layer has common registration and the layers are held together with registration pins to combine the layers. The final layer combination is photographed for map reproduction. The principal reason for doing this, even in black and white mapping, is safety; if you spill ink on one layer, you do not have to clean or redraft the entire map, only that layer. Layering is also part of color map production. The digital GIS layer is an adaptation of the analogue notion of layering in multicolor map production. To produce multiple colors on paper, the information needs to be prepared in layers so that the printing plates for the different colors can be made. To produce the 7.5-minute topographic map, the USGS cartographers need to produce the following layers:

- *Black*. Most artificial features except large roads. These include smaller roads, buildings, political boundaries, place and feature names, and benchmarks.
- *Red*. Built-up urban areas (screened to pink) and larger roads.
- *Green*. Continuous tree cover.
- *Brown*. Topography.
- *Blue*. Hydrography (water features and symbols).
- *Purple*. Revisions from aerial photographs not yet field checked.

Drafting each of these feature sets in black ink on a different layer of acetate, using these acetate layers to produce printing plates and then applying different colors of ink to the plates produces a multicolor map containing a high density of information. This separation of the world into layers was also adopted by the early (and current) versions of CAD, where features would be placed in layers not only on the basis of their color but also their symbolization on the final drawing. CAD layering is extensive; a single drawing may consist of dozens of layers, and individual features may be made up of elements from multiple layers. In a GIS layering system the entire feature is usually stored on a layer, and the result is many fewer layers to represent the real world.

The notion of structuring the data in layers has a lot of advantages and a long tradition in GIS. In fact, it was the ease with which digital layers of suitability criteria could be combined that served as the impetus for the early development of raster GIS to replace the cumbersome methodology of analogue overlay popularized by Ian McHarg (McHarg 1969). On a topographic map, it is difficult to examine the drainage pattern revealed by the blue lines representing streams because the other information obscures what you want to see. If that layer is represented on a computer screen in a GIS, though, you can simply turn off the other layers, and the stream layer is more visible. You can also control the order in which layers display on the screen by dragging them to the bottom or top of the list of available layers. The drawing order on a paper map is determined during the printing process, and you cannot vary what layer overlays what other layers. Whether you are drawing geographic information on a computer screen or printing it on a map, you are drawing on a blank surface, and you will be adding information in layers that draw in a specified order. At the level of display, all data are presented as layers in a GIS whether or not it is stored as layers in the database.

But the question here is one of representation in the database, not display on the screen. The idea of object-oriented databases developed in the 1990s out of object-oriented programming languages. This style of programming has mostly replaced the traditional style of sequential programming where programming code was long and involved, containing many calls to subroutines, program libraries, and so on. In object-oriented programming systems (OOPS) the software engineers write code in small snippets. Code can be associated with an object, a window, a dialogue box, or some other type of object defined in the system. This object then possesses some "intelligence" and can communicate and direct activities of other objects if this is allowed within the programming system.

The idea of extending object orientation to databases was a logical outgrowth of this programming style. You can conceptualize features in a database as objects, give them properties, and allow them to communicate with other objects in the database. This functionality elevates the feature from being something passive in the database that you can retrieve, manipulate, and replace in the database to being an object with properties of its own and a set of functions it can perform on itself or other objects. Object-oriented geographic databases are becoming more common in GIS, and each software company has its own terminology for them. Right now many organizations are taking their legacy-layered databases and converting them to object-oriented databases. To be consistent, we are going to refer to this type of database as a spatially enabled relational database system (SERD).

Modeling geographic features as objects in a SERD requires that the database adhere to the general principals of object orientation:

- *Hierarchical organization.* Objects usually belong to a hierarchy, and the position in the hierarchy is an important property of the object. Objects higher in the hierarchy own objects below them, and lower objects belong to higher-level objects. So a building may belong to one or more land parcels depending on whether it crosses the boundary of the parcel, and a transformer may belong to zero or one utility pole but not to two poles. This notion of hierarchies is central to the programming where the highest level might be something called a project; other objects, which are different views into data, belong to the project; and layers or sets of features belong to the data views. The project then only owns the layers through the data views. Organizing the geographic features as objects implies that there is some kind of hierarchy of objects (i.e., that some objects control objects below them in the hierarchy).

- *Objects have properties.* This characteristic of object-oriented databases is not much different from feature-oriented databases. In both layer and object geographic databases, a feature must have the property of type. For example, do we represent this feature as a number or text? If it is a text feature, it may have to have a property of length (number of characters). If it is a numeric feature, how many places to the right of the decimal point does the system have to store? Object properties are often stated in the form of questions such as Can this object be modified once it is created?, Is this property of this object mandatory to be specified or can it be left out?, or Is there a particular set of values or range of values that this property may assume?

A key property that an object has in a geographic database is its location. Location may be modifiable or not, depending on how you set up your database. A land parcel has a location which cannot be modified. If you change the locational property of a land parcel by combining it with another parcel, you have actually destroyed the original parcel object and created a new one with a nonmodifiable locational property. But a building is movable and can change its location, although this is not a common occurrence. One of the important features of designing databases is that if something occurs very rarely but does occur, it needs to be allowed for in all objects. You could not decide that some buildings are movable and others are not. In an electric utility geographic database, a transformer is an object that is movable, but it may have the property that it must belong to a utility pole, a substation, or a staging yard. Its absolute location, latitude, and longitude will change as it is moved from a staging yard to a utility pole, and its relationship with objects will as well. First it belongs to a particular staging yard and then it belongs to a particular utility pole, actually sharing the locational property of the object to which it belongs. This notion that location is just another property of an object in the database means that the geographic information is now stored directly with the nonlocational properties, traditionally referred to as the attribute information. This means that deletion or insertion of the object into the database is complete when it occurs.

And if that object is linked to other objects, changes will occur there as well. To continue the transformer-utility pole example, if you remove a pole from the database, you will have to remove any objects that belong to that pole as well, and the system will force this operation. In a layered database where the poles are on one layer and the transformers another, the user would have to remember to do it, and forgetting could result in a display of transformers appearing to hang in space unattached to a pole. In a layered geographic database it is common for a feature to have two types of information: geographic information (the where of the feature) and attribute information (the what of the feature). In an object-oriented geographic database this information is not separate but is simply another piece of information we have about the object. It may be actually stored in separate tables from the attribute information, but logically it is integrated data. In a SERD there is no distinction between the spatial and attribute information; it is all information we know about the object.

◆ *Object properties can be geographically conditioned.* An object may have different properties based upon where it is absolutely or relatively. A water control valve that is sitting on the shelf in an inventory warehouse controls no water flow, but that valve directly attached to a water pipe does have control properties over the downstream pipe. That property is made available by the change in location. In a national distribution system you may have a class of objects called metropolitan regions. When routing trucks around the country at one spatial scale, the locations of these regions can be represented as points in geographic space and a node on a transportation network. When the truck gets to the metropolitan region and the scale of analysis changes, the metropolitan region needs to be modeled as set of connected lines; a network within a network. Another example would be to display property parcels as points within a certain scale range, but after you zoom into a tighter, large scale, the parcels would show as polygon features. At all scales, though, you would have access to the same information. In a layered database you would need to have separate layers for the feature as points and polygons.

◆ *Relationships between objects have properties.* The idea that objects have properties was part of database design from the start, but object-oriented databases have added the idea that objects can relate to other objects and that these relationships themselves are objects of a sort, so they can have properties. An object may have a mandatory ownership property relationship. An example in an object-oriented GIS would be that a building object must belong to one or more land parcel objects. You would have to define this as one building to one or many land parcels because of the possibility of a building straddling a parcel line. This relationship property could be defined so that the addition or removal of a building object from the database would trigger a change in the properties of the land parcel(s) it belonged to (i.e., the properties of the number of buildings on this parcel). A valve object in a water utility GIS could be related to a pipe segment downstream in such a way that when the property of the valve is closed, water can no longer flow into the receiving pipe when modeling water flow in the system. The ability to define relationships between objects is what makes it possible to associate behavior with objects. Object properties can act as triggers so that when the property is set to, say, true, some behavior is triggered in the object. That behavior could be something as simple as changing the symbol on the display when the object is clicked on. A maintenance crew with a global positioning system (GPS) that sends the real-time location to a dispatcher could change status from working and not available to available for work and the symbol on the dispatcher's display would change color, indicating availability for any nearby service calls. Or the behavior triggered by property states for objects could be more complex such as the sudden decrease in monitored pressure in a water main triggering the closing of the immediately upstream and downstream valves in the distribution system through a supervisory control and data acquisition (SCADA) system linked to a water utility GIS.

◆ *Objects can belong to classes.* When you consider the number and variety of objects that are in the real world, you realize what a challenging task it is to design good geographic databases to represent this real world. But fortunately there are usually logical groups of objects in which you are interested. A forestry company might take the standard object of a tree and create three classes of trees: trees that are good for pulp, dimension lumber, and all other trees. A distribution company might group the object we call a truck into a number of truck classes such as refrigerated, liquid bulk, dry bulk, and general cargo trucks. All these objects would have their own properties, but they all belong to the superclass of object we call truck, which is different from the object class we call automobile. Both these object classes belong to a higher class of objects we might call road vehicles. All subclasses share the properties and behaviors of the superclass.

◆ *Properties can be inherited.* Combining the notions of hierarchical structure and properties, object-oriented systems are designed so that subclasses of objects inherit the properties of the superclass to which they belong. In the example of road vehicles all these classes of objects have a property of number of wheels and number of possible passengers and a property of a driver. That property is not mandatory — a road vehicle can exist without a driver — but if it has a driver, it can have only one driver at a time. It could have multiple passengers, however. Each of the subclasses such as truck or automobile inherits these properties from the superclass of road vehicle in addition to having its own properties.

Making the mental switch to thinking of real-world things as features in a layer of a geographic database to objects with properties, behavior, and relationships with other objects is difficult because the notion of layers goes so far back in cartography, printing, and CAD. Many users will have no need for the additional capabilities that object-oriented databases provide. But through the same process that object-oriented programming languages have almost completely replaced sequential programming systems, object-oriented geographic databases will eventually replace the layer-based databases. For those people having difficulty with the transition, however, it is possible to design a geographic database within an object-oriented system that looks and behaves exactly like a layered database. Going the other way, setting up an object-oriented database in a layer-based GIS is not possible.

SERDs can be simple or complex; you do not need to set up relationships between objects nor must you program behavior into your objects. You can set up a simple database that replicates a layered database and, over time, define relationships and behaviors to make the data more active. Object orientation within the database as well as within the programming system, however, has allowed GISs to model the real world more closely. In applications where the GIS has to support the actual manipulation of real-world features in real-time, these systems are very useful. If a valve can be turned on or off in the GIS and can control the flow of virtual water in the downstream distribution lines, you need only to add a communication link between the real valve and the GIS, and by changing the property of the valve in the GIS from open to closed, you can change the property of the real valve as well. If a segment in a metropolitan transportation system is suddenly closed due to an accident, you can identify all planned routes of delivery trucks that were supposed to cross that link in the next 2 hours and send a beeper message to all drivers to reroute around the closed link. The changing of a property of a link from open to closed can trigger that message automatically if the system is set up that way. Over time all GISs will adopt object-oriented programming and databases, and the only layer-based systems will be those whose sole functions are to produce mapped output either to a computer screen or paper.

Construction of GIS systems has been going through radical changes over the last few years. Traditional GIS deployments have been constructed using a combination of spatial and attribute data that have been stored in separate file formats and brought together through the process of linking the spatial feature to the attributes by some sort of a unique identifier, the so-called geo-relational model. Spatial data has been stored in varying proprietary file formats, and the attribute files have typically been stored in database or relational database formats such as DBF (dBase) files, MS-Access tables, or Oracle/SQL tables. We have discussed this issue previously when considering the geo-relational and SERD models for organizing your spatial data, but it is at the implementation stage where the final decision needs to be made.

One of the leading GIS vendors, Environmental Systems Research Institute (ESRI), has traditionally used a file format known as the coverage. In more recent years, with the move toward standards and interoperability, it introduced a file format called a shape file, which is a fully documented and a nonproprietary format for the spatial data, and the attribute data are stored still in a DBF format. In very recent years the company developed a new format called the Geodatabase, its term for a SERD, in which it stores both the spatial and attribute data in the database itself. This concept is not new. Intergraph has been using this type of data storage model for quite a few years, but its solutions haven't been marketed as widely, so it hasn't caught on as well as ESRI's solutions. Oracle introduced a technology called Oracle Spatial a number of years ago that also used this same methodology of storing both spatial and attribute data in the same workspace. This caught on very well in large organizations, but because of the relatively high cost of its products, it also hasn't been used as widely. Other database software developers have also produced spatial data solutions that would fall into the category of a SERD.

Based on trends that have been seem in these recent years, the concept of storing all of the data in one place seems to be here to stay, at least until the next new revolutionary concept is developed. Nonproprietary, easily accessible formats like the shape file will probably exist as well and will allow for easier exchange of data between different systems and platforms.

The decision of whether to use layers or an object-oriented SERD is an important because it guides the design process. If you are building a set of independent layers, each layer is its own thing and can be designed separately. If you are building an object-oriented SERD, you need to be aware of relationships between features classes and explicitly describe those relationships; it forces you to think of the database as a single entity rather than a set of coincident layers. The shift from layers to SERDs will take some time, especially for those who are used to working with layers. True objects, if carefully designed and constructed, represent the real-world features much more closely than features in a layer, and any database model that more closely approximates the world it abstracts is a better model.

Representing Geographic Features

In introductory classes students are always taught that there are three classes of geographic vector features: points, line, and polygons. In advanced classes they learn that there are many more types of geographic features and that not all geographic information systems support all types. Certain types of geographic features are extremely useful in some applications and completely useless for others. A full topologic linear transportation network is essential for a GIS to support physical distribution but absolutely useless for one that is set up to facilitate the legal transfer of land. Both systems are GISs, but their model of the world is quite different. So it is important to understand the geographic nature of the features you need to represent in your GIS so that you can acquire the right kind of software to model these features.

In the 1980s, as different software manufacturers started programming for new types of features, the U.S. federal government became concerned because different organizations within the government were purchasing different GISs, and they were having trouble exchanging data. So in 1980, the USGS was designated the lead agency for dealing with this problem with respect to spatial data, and they initiated a process which resulted 12 years later in the publication of the standard for the transfer and exchange of spatial data, the Spatial Data Transfer Standard, or SDTS (FIPSPUB 173-1, 1994.) The most recent version was published in 1998

(ANSI NCITS 320-1998), and there will probably be additional changes as users come up with different representations of real-world features that need to be exchanged between GISs.

It was the need to exchange and transfer real-world features that forced the creators of the standard to think very clearly about exactly how to represent them from the top of the process (the data model) to the bottom (the physical file structure on the computer). Although the many vendors of GIS software often do not use the same terminology as SDTS, each of their supported data types — whatever they call them — must conform to one of the SDTS data types if they wish to sell systems to the federal government, a large market to be sure. The acceptance of these standards has not lived up to expectations, and creative people in the GIS world are regularly developing new ways to represent geographic features. We use SDTS as the foundation for our discussion principally because different GISs implement different features and have their own internal names for these features. What SDTS calls an arc is not what a particular GIS might call an arc. SDTS at least provides a common language for discussing types of geographic features. There are two profiles in the standard, one for raster or grid cell data and one for topologic vector data. The discussion that follows is for the topologic vector profile. If you wish to understand the raster profile, you can find it on the SDTS Web site, mcmcweb.er.usgs.gov/sdts/.

Topologic Relationships

Topology is one of the most confusing and difficult to understand concepts in the design of spatial data sets partly because people often confuse it with topography. They are completely different things. If a data set has topology, it means that relationships of adjacency (for two-dimensional features) or connectivity (connections between linear or one-dimensional features) are explicitly stored in the database. Geographers often make a distinction between absolute and relative location. The absolute location of a feature places it in a two- or three-dimensional coordinate space (e.g., latitude, longitude, and altitude). The relative location is that feature's location relative to other features (e.g., which other features are adjacent to it). It is possible to represent the absolute location of a feature in a digital file with an ordered listing of the number pairs (x/longitude and y/latitude) that describe its location in a coordinate space. This is sometimes referred to as spaghetti vector data because, like a plate of cooked spaghetti, the lines may overlay each other, but the system has no way of knowing that they do because each line is stored independently of the other lines; whether a line crosses another is not stored and therefore not known to the system. You, with your birds eye view, can see that the two lines intersect, but the software does not have that view of the data; it sees the line only as a series of points in a file.

For topologic vector data, on the other hand, you must explicitly store the relationships of adjacency or connectivity or be able to construct them as needed. Without the knowledge of adjacency and connectivity, you cannot execute spatial queries. It is possible to query features based on their attributes or properties (i.e., display all the sales territories that had sales increases over 20 percent) but not possible to locate all the sales territories adjacent to one particular territory to aggregate their information. Spatial querying is at the heart of what most users do with their GIS, so you must have a data model that allows it to occur. In the early days of GIS when computer central processing units (CPUs) were small and slow, topology had to be stored explicitly in tables in the database, and spatial queries were executed by making many hits, or reads, into those tables. As the processors became more powerful and faster, it was no longer necessary to store the topology in tables. Now we can execute the spatial queries through sophisticated indexing and searching procedures. The systems moved from complex data structures and simple queries to simple data structures and complex queries, and the user sees the same result.

Polygon topology is easier to visualize than to describe. Figure 4.4 shows two-polygon data sets, one with topology and one without. The nontopologic data set stores only the coordinates of the points where the line that outlines the polygon changes direction. The first pair of points in each polygon is duplicated at the end to close the polygon. In this data structure the polygon that appears to be where the two overlap is not in the database. The bottom part of Figure 4.4 shows the same polygons in a topologic structure, and it demonstrates the additional database complexity that topology requires. A single file or table can hold the nontopologic information, but storing topology requires a linked set of several data tables. The key table in the set of tables shown here is the Chain_node table (often referred to as the Arc_Node table, but we are using the terminology of the SDTS); it is here that the topology exists. Each chain has a specified node it comes from and one it goes to. This produces an implied direction, and that makes it possible to identify the polygon to the right and left of the chain. The G-polygon table contains the identifiers of the chains that make it up. These chains are listed in clockwise order, and if a chain needs to be reversed to keep the clockwise order, a negative sign is added. So, for example, G-polygon B is made up of chains 3 and 4, but to get the chains to go in clockwise order, they need to be reversed (see Table 4.4 for a discussion of G-polygons). So it does not matter what the implied direction of the chain is, but it does have to have one. This is determined by specifying which is the from and which the to node. The node and chain tables do not contain topology, only the coordinates that describe the location of the nodes and chains.

For topology to exist in a G or GT-polygon data layer two conditions need to be met:

- There must be a universe or background polygon. Because every chain making up the G-polygons must have a right and left polygon defined, their needs to be a polygon to the right or left of polygons that do not touch any other polygon. The universe polygon provides for that. No feature in the database may extend beyond this universe polygon, so correctly choosing the extent of that polygon is an important early step. Fortunately, if your area of interest extends beyond the current universe polygon, you can create a new one and rebuild the topology.

Nontopologic (spaghetti) Structure

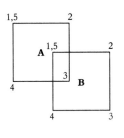

Polygon A

ID	X	Y
1	0.00	2.00
2	1.25	2.00
3	1.25	1.00
4	0.00	1.00
5	0.00	1.50

Polygon B

1	1.00	1.50
2	2.00	2.50
3	2.00	0.00
4	1.00	0.00
5	1.00	1.50

Key

X Y

1 ID value and coordinate pairs mark change of direction or closure of polygon. Notice that the polygon ends with a duplicate of the first point; this closes the polygon.

A polygon identifier

Topologic Structure

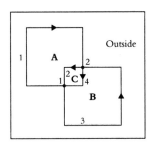

Key

A polygon identifier

1 line segments defining polygons.

2 nodes marking the beginning and ending of line segments.

▶ direcion of line segment.

- negative sign in chain list means direction of chain must be reversed to create polygon in clockwise direction.

Node Table

ID	X	Y
1	1.00	1.00
2	1.25	1.50

Chain Node Table

ID	from_node	to_node	left_poly	right_poly
1	1	2	Outside	A
2	2	1	C	A
3	1	2	B	Outside
4	2	1	B	C

Chain Table

ID	X	Y
1	1.00	1.00
	0.00	1.00
	0.00	2.00
	1.25	2.00
2	1.25	1.50
	1.00	1.50
	1.00	1.00
3	1.10	1.10
	1.00	0.00
	2.00	1.50
	1.25	1.50
4	1.25	1.50
	1.00	1.50
	1.00	1.00

G Polygon Table

ID	Chain_list
A	1, 2
B	-3, -4
C	4, -2

Figure 4.4 Topologic and nontopologic structure: polygons.

- Where ever a chain touches another chain. there must be a node, and that node will be the end point of one chain and the beginning point of the second (or third, fourth, etc.). You will notice that this is not the case in the spaghetti vector example at the top of Figure 4.5, and this is one of the most important requirements to construct topology. A GIS cannot create topology if there are any places in the database where a chain crosses another and there is no node. It is possible to have the system automatically find these locations and place nodes there, however.

So what is the advantage of this complex structure? The principal advantage is that many geographic queries and processes can be conducted using only the Chain_node table. The actual coordinates defining the features are only needed for drawing the results on the screen or a map. As an example, say you wanted to remove G-polygon C from the layer shown in the bottom of Figure 4.5. The system would do this by searching the Chain_node table for all chains whose left or right G-polygon is C (chains 2 and 4), removing them from the appropriate tables, and then reconstructing the Chain_node table for the new topology, which would now have only one G-polygon made up of chains 1 and 3. The following common geographic functions on polygons all require knowledge of the topologic relationships between polygons:

- *Clipping out a subset of polygons with another polygon.* Say you acquired a data layer of all the land parcels in a large region, but you are only interested in the parcels within a particular area. You create a polygon layer of this area of interest and use it like a cookie cutter to clip out the larger parcel layer, keeping only those polygons inside the area of interest.

- *Overlay of polygon layers.* Perhaps you have one layer of wetlands and another of watershed boundaries and you want to know how many acres of wetland are in each watershed. By overlaying the wetland layer with the watershed layer, you can attach the watershed identification to the wetlands inside it. And if a wetland has some peculiar topography so that it drains in two directions into different watersheds, the overlay process will create two wetland/watershed polygons correctly reflecting the situation.

- *Dissolving lines between adjacent polygons that are similar on some characteristic.* If you had a parcel layer of polygons and information on the principal land use for each parcel, you could use a dissolving function, which would remove the lines between adjacent parcels that had the same principal land use. The result would be a much-reduced set of polygons that showed principal land use.

Vector GISs that do not have the ability to store or create polygon topology are useful only for map display and have relatively few analytical capabilities; therefore, vector topology is almost standard in GISs today.

Nontopologic (spaghetti) Structure

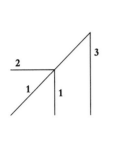

Chain 1

x	y
0	0
1	1
1	0

Chain 2

x	y
0	1
1	1

Chain 3

x	y
1	1
2	2
2	0

Key

1 **x y**

ID value for chain; **x y** pairs for beginning and end of chain and at each change of direction.

Topologic Structure

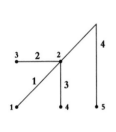

Chain_Node Table

Line_Id	From_Node	To_Node
1	1	2
2	3	2
3	2	4
4	2	5

Node_Table

Node_Id	x	y
1	0	0
2	1	1
3	0	1
4	1	0
5	2	0

Notice: Nodes exist at beginning and ends of chains. Where chains touch, there must be a node. A chain is related to only two nodes, a node it originates *from* and a node it goes *to*.

Line_Table

Line_Id	N_XY_pairs	XY_pairs
1	2	0,0,1,1
2	2	1,0,1,1
3	2	1,1,1,0
4	3	1,1,2,2,2,0

Figure 4.5 Topologic and nontopologic structure: lines.

It is also possible and useful to create topology for linear features. Although polygon topology requires knowledge of adjacency (which polygons are to the right and left of each chain defining them), line topology requires knowledge of connectivity (which lines touch which other lines). Nontopologic, or spaghetti vector, line segments do not carry this information, only the coordinate pairs or mathematical descriptions of the arcs. Topologic linear data layers must have a node where lines touch each other, and the Chain_node table relates the line segments or arcs to their beginning and ending nodes (Table 4.5). The value of this kind of topology is principally in network analysis. Graph theory, a branch of mathematics that deals with topologic networks, actually predates GIS, and within the last 20 years or so GISs have been incorporating the analytical functions that network topology makes possible. These are some common functions you can perform if your data layer has linear topology:

- *Find the shortest path between two locations.* So long as the locations are nodes on the network, there are algorithms that will seek out the shortest distance in time or distance between those two points across the network.

- *Location-allocation.* Given a set of demand points (e.g., grocery stores) and a set of supply points (e.g., warehouses), assign each grocery store to the nearest warehouse in the network. Location-allocation is an important model in operations research (OR), an area of applied mathematics, and some of the spatial modeling functions developed by OR practitioners have been incorporated into GIS.

- *Find an alternate route.* If a segment of a network is blocked or unavailable, you can recalculate the new shortest route around the blockage.

- *Create an optimum route to multiple nodes.* Known as the traveling salesperson problem, it is commonly used for school bus and garbage pickup routing. This functionality in a network GIS will determine for you the least-cost path through a network from a given starting point to a given end point, stopping at a set of selected nodes in the network.

Because of the obvious advantages of the topologic vector data model, it has become standard in GIS. When computers were much less powerful than those of today, the Chain_node polygon topology structure was needed because the computers could not hold large numbers of features in random access memory (RAM) at one time. With the explosion of computer memory and computation speed, however, it is now possible to store vector information in the spaghetti vector data model and create topology as needed in the computer's memory but not in the database. This construction of on on-the the-fly topology is becoming more common in GIS and has several advantages. When topology is stored explicitly in data tables, the removal of a single feature requires the reconstruction of the topology of the entire layer of information. This process can consume significant time, and there is also the danger of losing some key information on relationships between tables, which destroys the topology as well. The shift from layered GIS databases to object-oriented databases has also driven the move away from explicit topology maintained in the database to on-the-fly topology constructed when and where needed. In a layered database the spatial data, the coordinates representing the location of the features, are usually kept in a set of tables separate from the attribute information, what we know about the feature.

This geo-relational data structure for the topologic vector data model has been around for many years but is not well suited for the object-oriented data structure. In object-oriented GIS databases, the spatial and attribute information is stored together in a single record for each feature in the database. This simplicity of storage means it is harder to lose features and easier to completely insert and remove features from the database without having to reconstruct the topology every time. To use an analogy of building a house, with the geo-relational, the explicit topology structure to add a room to the house involves reconstructing the entire house after you put the new room on. But the system knows exactly what other rooms the new room is next to. In the implicit topology, object-oriented database, you can add the room to the house without affecting the other rooms. When it becomes necessary to know which rooms are next to the new room, the system can process the data and discover only that piece of information. Generally, object-oriented implicit topologic databases allow you to deal with the pieces without affecting the entire database, but the geo-relational data structure means you have to deal with the entire database every time. As databases get very large, this clearly becomes an issue of processing capability and speed.

The type of topology we discussed is implemented in the database through sets of related tables, hard-wired into the data. ESRI's rules-based topology, the geodatabase, allows users to specify simple or complex topologic relationships for single- or between-feature layers. A simple rule for G-polygons might insist that there be no line segments dangling in the data set. A feature that violated that rule would be flagged or identified and the user prompted to fix it or allow an exception to the rule. A topologic relationship between layers might require that all features from one layer (e.g., buildings) lie completely inside a single feature of another layer (e.g., land parcels). Buildings that intersected more than one land parcel would be flagged for correction or exception. Within a network there could be a rule that all line segments be linked to at least one other line segment (i.e., there can be no segments that are not part of the network). Rules-based topology can get complex; there are 25 possible rules that can be applied in the first version of ESRI's topology and more are planned. Unlike arc-node topology, this type of topology is not directly stored in the database. Rather, the rules are evaluated on the fly during editing when the user expressly requests the rules to be implemented.

Types of Spatial Objects

The creators of SDTS chose to use a classification system based on the number of spatial dimensions the object requires in the database. There is a hierarchical component to this structure as well; you cannot describe a line without knowing the points that make up at least the beginning point of the line and some way to describe where the line goes from there. One-dimensional lines, which have points as their beginnings and ends, bound two-dimensional features, and everywhere they change direction. A GIS is an abstraction of reality, and one of the most important abstractions you have to make is do decide what kind of digital feature, or SDTS data type, you are going to associate with the real-world features you must deal with:

- *Zero dimensional features* (see Table 4.2). These are point features with no length or breadth and therefore appear to take up no space. They may, however, represent a real-world feature that has dimensionality. For example, a point representing a well is an NE point according to this classification scheme, but a well is really a cylinder or possibly a more complex three-dimensional object that changes direction and width at different depths. The well is represented as a point feature principally because that it how it is represented on a map, but if your GIS needed to model in three dimensions, you would need to model it differently. Some of the point entity types defined in SDTS are for cartographic convenience, such as label and area points and do not represent any real-world feature.

- *One-dimensional features* (see Table 4.3). Linear features have length but no breadth. They may represent features with breadth that can be stored as a property of the feature or object. The transfer standard explicitly includes linear features in networks and distinguishes between networks that are planar (fit entirely in two dimensions) and nonplanar (require more than two dimensions to work). In a planar network if two lines cross, a node must exist. but this is not a requirement of nonplanar networks. Most transportation systems are nonplanar networks with underpasses, over-passes, and tunnels that require three dimensions, but there are ways to model them as planar networks, which are simpler computationally. The network model of reality is not discussed in this book

- *Two-dimensional features*. These are areal or polygon features that have breadth as well as height (see Table 4.4). These features are more com-plex, and their verbal and graphic descriptions demonstrate this. The profile for vector data in the SDTS is for topologic vector data, the most useful data model for vector data, but data without topology can also be described within this standard.

What happens with real-world features that seem to have elements of more than one SDTS data type? Usually the response is to model it in different ways for different purposes. This is necessary even if the database is object-oriented, but the objects can be related to each other so that when a centerline is added or removed, the necessary changes to the curb and pavement area objects must be made as well. An object-oriented database would also allow for relationships between the street objects and the infrastructure on, below, and above the street. The standard also allows for composite features, which are groupings of related features of the same type. In transportation planning, for example, a route is defined as a set of linked network chains (LW). A well field for a water utility may be a composite feature made up of several entity points (NE). Not all GISs can handle composite features, however, so this becomes an important design issue.

Table 4.2 Spatial Data Transfer Standard Point Features

Point Features (Zero dimensional)

SDTS Name	SDTS Description	SDTS Abbreviation	Figure	Explanation
Point	A zero-dimensional object that specifies geometric location.	NP	●	The generic point. Line segments connect two points. No attribute information is attached to this type of feature; they simply represent geometry.
Entity point	A point used to identify the location of point features (or areal features collapsed to a point), such as towers, buoys, buildings, places, etc.	NE		Real-world features best represented by points at a specified scale. A point at one scale but a polygon at another scale may best represent a feature (e.g., place).
Label point	A reference point used for displaying map and chart text (e.g., feature names) to assist in feature identification.	NL		Not a real-world feature; to place labels on features or text on maps.
Area point	A representative point within an area usually carrying attribute information about that area.	NA	✳	Usually the centroid (approximate center) of a polygon. Often is the default label point.
Node	Topological junction of two or more links or chains.	NO (planar graph)		An intersection joining two segments (links) of street.

Table 4.3 Spatial Data Transfer Standard Linear Features

Line Features (One Dimensional)

SDTS Name	SDTS Description	SDTS Abbreviation	Figure	Explanation
Line segment	A direct line between two points (NP).			The simplest linear feature, a straight line connecting two points.
String	Connected nonbranching sequence of line segments specified as the ordered sequence of points between those line segments. Note: a string may intersect itself or another string.	LS		This is the so-called spaghetti vector data structure. Line segments have no official beginning or end, and if they intersect, a point (node) is not defined. Mostly used for mapping only.
Arc	A locus of points that forms a curve that is defined by a mathematical expression.	AC (circular arc), AE (elliptical arc), AU(uniform b-spline), AB (piecewise bezier)		To represent complex linear features, often artificial. Some GISs cannot model this type of spatial feature and will convert them to strings. A circle as an AC-type feature is described by a point and a radius; a circle as a string is a set of a large number (often 256) strings.
Link	Topological connection between two nodes. A link may be directed by ordering its nodes.	LQ		For modeling networks. A network chain is a more complex representation of a link in a network.
Chain	A directed nonbranching sequence of nonintersecting line segments and (or) arcs bounded by nodes, not necessarily distinct, at each end.			Most common feature in topologic vector data layers.

Special Cases of Chains

SDTS Name	SDTS Description	SDTS Abbreviation	Figure	Explanation
Complete chain	A chain that explicitly references left and right polygons and start and end nodes. It is a component of a two-dimensional manifold.	LE		A part of arc–node topology.
Area chain	A chain that explicitly references left and right polygons and not start and end nodes. It is a component of a two-dimensional manifold.	LL		A part of arc–node topology.
Network chain	A chain that explicitly references start and end nodes and not left and right polygons. It is a component of a network.	LW (planar network), LY (non-planar network)		
Ring	A sequence of nonintersecting chains or strings and (or) arcs, with closure. A ring represents a closed boundary but not the interior areas inside the closed boundary.			
G-ring	A ring created from strings and (or) arcs.	RS (ring of strings), RA (ring of arcs, RM (ring of mixed composition)		
GT-ring	A ring created from complete and (or) area chains.	RU		

95

How They Did It – Selecting Software to Support Dynamic Segmentation

People who are explaining GIS to novices quickly learn to discuss points, lines, and polygons as the fundamental features of a GIS. But as the tables in the section on the SDTS demonstrate, there are additional types of features that can be modeled in a GIS. Of particular importance to transportation planners and engineers is the feature known as a route. A route is a (usually) connected set of lines that make up a single management unit (e.g., PA Route 34). Although the GIS needs to manage all the line segments that make up that route, it is the route that is the principal management unit. It is the feature that gets upgraded, has other facilities and features associated with it, and has been the principal geographic feature for transportation planning and engineering since well before GIS.

Secondly, this group of professionals measures location differently from others. Most GISs rely on latitude and longitude or some projected coordinate system to measure the where in the system, but in applications using principally linear features, location along the route or feature of interest is a simpler, and traditional, way of locating things; this is called linear referencing. A feature, such as a sign, exists at some distance along a route that has a known beginning point. So a location such as mile 5.432 on Route 34 is as meaningful and locates the sign as well as the x/y coordinate pair that also locates it. So instead of a point being located with its x/y values, it has x, y, and m values where m represents the distance along the route. If a GIS supports dynamic segmentation, it is possible to attach multiple sets of attributes to any section of a route. Linear features along the route such as sections of pavement for management purposes can be described in a table with a beginning and ending location along the route. Within that same section of route there will be other features like signs that have their route location stored as a single point value along the route. You can also specify side of the route and offset so that it will plot in its correct location. It is easy to see how important this data structure is to organizations that deal with transportation construction and planning and any organization with interests that are principally linear rather than areal in nature.

For this reason, many transportation planning and engineering installations have selected software systems that support this data structure. As Eric Solomon of the Transportation and Civil Engineering Division of provincial government in Alberta wrote to one of the authors in email, "Our software was selected through a comparison of products (Intergraph, ESRI, and AutoDesk Map Guide) in 1997. The Department was a long time client of Intergraph, and it was more of an informal process to validate that we were using the best vendor. Since we had a legacy of Intergraph data (Microstation, MGE, IPLOT), and since Intergraph still seemed to best support dynamic segmentation, we stayed with what we had."

Table 4.4 Spatial Data Transfer Standard Polygon Features

POLYGON FEATURES *(Two dimensional)*

SDTS Name	SDTS Description	SDTS Abbreviation	Figure	Explanation
Interior area	An area not including its boundary.			
G-polygon	An area consisting of an interior area, one outer G-ring, and zero or more nonintersecting non-nested inner G-rings. No ring, inner or outer, must be collinear with or intersect any other ring of the same G-polygon.	PG		Allows for donut polygons with holes inside; the rings may not touch each other, and there may not be an area inside the hole, an island (hole) in a lake (G-polygon) with a pond on the island.
GT-polygon	An area that is an atomic two-dimensional component of one and only one two-dimensional manifold. The boundary of a GT-polygon may be defined by GT-rings created from its bounding chains. A GT-polygon may also be associated with its chains (either the bounding set or the complete set) by direct reference to these chains. The complete set of chains associated with a GT-polygon may also be found by examining the polygon references on the chains.	PR (made of rings), PC (made of chains)		The polygon defined by arc-node topology. Does not allow for interior polygons (see PG- polygon).
Universe polygon	Defines the part of the universe that is outside the perimeter of the area covered by other GT-polygons (covered area) and completes the two-dimensional manifold. This polygon completes the adjacency relationships of the perimeter links. One or more inner rings and no outer ring represent the boundary of the universe polygon. Attribution of the universe polygon may not exist or may be substantially different from the attribution of the covered area.	PU (made of rings), PW (made of chains)		This is sometimes known as the "outside" polygon and must exist for adjacency relationships to be defined for the rest of the polygons that make up the covered area. It usually has no attribute information but may. Some GISs recognize it but do not display it or list it in tables.
Void polygon	Defines part of the two-dimensional manifold that is bounded by other GT-polygons but otherwise has the same characteristics as the universe polygon. The geometry and topology of a void polygon are those of a GT-polygon. Attribution of a void polygon may not exist or may be substantially different from attribution of the covered area.	PV (made of rings), PX (made of chains)		Allows for intrusions of the universe polygon inside the area covered by GT-polygons.

Issues around the Third Dimension

Paper and computer screens are two dimensional and the world is three dimensional; incorporating the third dimension into GIS has been an ongoing struggle. GIS evolved partly out of the need to transform paper maps into digital form, and paper has only two usable dimensions. Cartographers have developed several conventions for representing the third dimension — elevation, topography, or terrain — on paper and computer screens. Most common is the use of contour lines (hypsography), a straight-down projection of the third dimension onto the two-dimensional paper surface. Requiring more craft and skill, shaded relief and analytical hill shading are ways to depict the third dimension on a paper map or computer screen. The key elements of the third dimension that sometimes need to be modeled in a GIS are as follows:

- *Elevation.* The height of a part of the real world above a chosen three-dimensional starting point (e.g., mean sea level).

- *Slope.* The change in elevation between two points over a given distance.

- *Aspect.* The compass direction a part of the real-world faces.

Incorporating these and more complex elements of the third dimension such as drainage basins and networks into a GIS is not an easy task, and an organization that is considering a GIS should expressly deal with the questions of whether and how to do this.

In the raster profile, SDTS briefly mentions a three-dimensional data object, the voxel. In addition to width and breadth the voxel has height. Using voxels to model reality is critical in many application areas. In the environmental field true three-dimensional systems are used in oceanography, meteorology, and climatology. And modeling in three dimensions at the scale of the human body is at the core of magnetic resonating imagery (MRI) and three-dimensional image holography. These are very specialized systems, and there is no implementation of the use of voxels in any commercially available system that calls itself a GIS. Part of the problem is the sheer size of the data sets. In a two-dimensional raster if you decrease the size of the pixels by half, say, from 10 to 5 m, or extend the study area a given distance in x and y, you increase the number of cells by 4 times. In a three-dimensional voxel model this is an eightfold increase. Even though computing speed and abilities are increasing rapidly, it takes very powerful computers to handle sizable three-dimensional voxel data sets.

The raster data model incorporates elements of the third dimension as values in a cell (see Figure 4.6). This is sometimes referred to as a 2.5-D model because the elevation is not directly part of the model but is stored as an attribute of a two-dimensional pixel. These models are also referred to as digital elevation models (DEMs) or digital terrain models (DTMs). The use of rasters to store and model elevation, slope, and aspect is one of the real strengths of the raster model. Some principally vector GISs directly incorporate the ability to handle the raster data model and treat the third dimension in this fashion, and some provide this functionality as add-on software components. So an important part of assessing needs is to understand exactly how the third dimension will be modeled, if at all.

Figure 4.6 Raster digital elevation model.

Figure 4.7 Analytical hill shading, digital elevation model.

A common use of the 2.5-D raster data model is analytical hill shading (see Figure 4.7). Having raster layers of slope and aspect of an area, it is possible to shine artificial light on the area from a certain direction and elevation above the horizon and create a pattern of light and shadow that provides a visual sense hills and valleys. Even the simple incorporation of a hill-shaded image as a backdrop can bring a sense of the third dimension into a GIS.

In the topologic vector data model, the easiest way to handle elevation is through the use of contour lines (see Figure 4.8). On maps the convention is to space contour lines at a constant elevation change such as 10 ft and to print the lines thicker every five or ten contour lines. Sometimes, but not always, these lines are printed in brown. This is the way the USGS represents elevation on two-dimensional maps, for instance. This is not difficult to implement in a GIS; the elevation of the contour

Designing Spatial Data

line is stored as an attribute, and contours can be selected or deselected using standard attribute queries. Many utilities and government agencies construct the core of their GIS database from digital aerial photography. If this photography involves stereo pairs of pictures, automated photogrammetric techniques can produce these contour lines with considerable detail and accuracy.

It is important to keep in mind, however, that contour lines like this make up very large data layers. This affects drawing speed and storage requirements. The viewer of a map can mentally ignore the contour lines on the map if the question to be answered dose not require their use, but the computer has no such luxury. But for many applications representing the third dimension in this map-like fashion is all that is necessary. It is also possible in most raster GISs to take vector contour line information with elevation attributes and derive a DEM from that with elevation, slope, and aspect represented in pixel form. Measuring slope or aspect directly from the vector contour lines is a more difficult process and usually requires additional programming.

The equivalent of 2.5 dimensions in the topologic vector data model is known as a triangulated irregular network (TIN; see Figure 4.9). TINs consist of a set of triangles known as Delaunay triangles that completely cover the area of interest. These triangles satisfy the condition that a circle drawn through the three nodes of the triangle will include no other node. The elevations of the corners of the triangles must be known. By constructing these triangles with known elevations, the

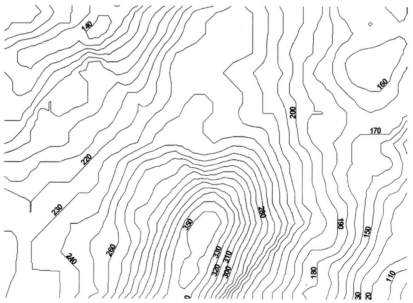

Figure 4.8 Contours derived from digital elevation model.

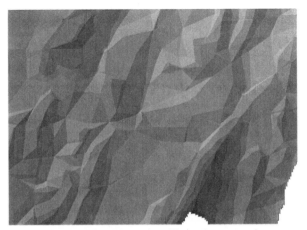

Figure 4.9   Triangulated irregular network.

calculation of slope and aspect is not difficult. These triangles are the vector equivalent of the raster DTM and are useful in the same way. TINSs are not easy to construct because they require knowledge of many points of elevation, three for each triangle, and there can be hundreds or thousands of triangles depending on the size of the area of interest and the level of detail necessary. For example, the sample TIN that was the source of Figure 4.9 has almost 6,000 triangles and over 4,000 data nodes. Generally, you need more triangles in areas of complex topography and fewer in simpler areas. The design issues for TINs are complex, and if you need to implement TINs, we suggest you consult the software documentation and the references at the end of this chapter. The true three-dimensional representation in the vector data model, the voxel equivalent, involves the three-dimensional extension of the Delaunay triangle, the tetrahedron. This model is important in modeling subsurface environments and in astrophysics and is beyond the scope of this book (Heller 1990; De Floriani et al 1984; Field and Smith 1991).

Another way that GISs have dealt with a three-dimensional world is in topologic networks. SDTS makes a distinction between planar and nonplanar networks. Planar networks are ones where crossing chains must have an intersecting node defined at that location. In nonplanar networks (i.e., more than two dimensional) networks chains are allowed to cross each other without creating a node. Because of underpasses, overpasses, and tunnels all transportation networks are nonplanar. But nonplanar networks are more difficult to handle in software, so programmers have developed a workaround to handle nonplanar networks as though they were planar. If a segment of divided highway goes over a section of road where there is no exit and therefore no way to move from the highway to the road, a planar network requires that a node be placed there. That node, however, will carry attributes in its table that describe the difficulty of making turns in that node. So if the right turn and left turn difficulty is maximum at that node, the only

Designing Spatial Data

option is to continue in the same direction of travel. The tricks to handling non-planar networks as planar are important and if your GIS must include networking capability, you need to understand these data structures well.

Table 4.5 Two-Point-Five- and Three-Dimensional Features

	Feature Type	Application Areas	Important Concerns
True three dimensional			
Raster data model	Voxel	Oceanography, climatology, meteorology, geology, medical imaging, physical anthropology	Large data requirements; not implemented in multipurpose GIS; requires separate software
Vector data model	TIN	Geology	Computationally complex, separate software require-ments often appli-cation specific, not implemented in standard GISs
Two–point–five dimensional			
Raster data model	DEM/DTM	Hydrology, geomorphology	Use as backdrop or need to model the third dimen-sion, ability to dis-play if not create raster backdrop images is common
Vector data model	TIN	Hydrology, geomorphology	Obtaining elevations of triangle corners, size, and number of triangles, dealing with variations in terrain complexity
Topologic network	Planar network	Transportation routing	Turning difficulties at pseudo-intersections

As the preceding discussions indicate, incorporating elevation and its associated derivatives of slope and aspect is not a simple matter in a GIS (see Table 4.5). The mathematics and programming necessary to accomplish this have been around for at least 15 or 20 years, and yet the difficulties of implementing even a 2.5-D data model in vector GISs have stymied many users. But it can be critical. For example, even calculations of distances traveled over a network vary depending on the underlying topography. The distance from point A to point B in a flat, or planar, environment is a simple application of the Pythagorean theorem, $C^2 = A^2 + B^2$. Calculating that distance over a complex terrain model is not nearly as simple. The effortless inclusion of the third dimension in vector GISs is probably a decade or more away. Until then, designers need to consider very carefully whether and how they will include the real-world elements of elevation, slope, and aspect.

Accuracy, Precision, and Completeness

Ensuring that the coordinates attached to the features in your GIS database are close to where they are in the real world is always a critical design concern. There are a number of factors that are important with respect to accuracy. The first is the positional accuracy of the data in the system. Positional accuracy is defined as the overall reliability of how close a feature is represented relative to its actual position on the surface of the earth. Positional accuracy is usually documented with a statement that a certain percentage (say 95 percent) of the visible features are within a certain distance (say, 2.5 ft) of the real-world location as determined by a more accurate method. Depending on the application of the GIS, the actual physical location of a feature is important to know when you need to make decisions relating a feature on one layer and a feature on another layer in the system. When designing, implementing, and using data in a system, you need to take great care to ensure that measurements and decisions made with the system are based on the layer with the lowest level of positional accuracy. A common error in the use of a system is to combine two layers with differing levels of positional accuracy and make measurements on the combined layer as though it were as positionally accurate as the most accurate, not the least accurate, source layer. Often the result can result in improper decisions.

More important to most GIS applications is the relative accuracy of the features in the system. The relative accuracy is defined as the overall reliability of how close a feature is to another feature on the same layer. It is possible to have a layer with a high level of relative accuracy but a low level of positional accuracy. For examples, features can be located using either conventional surveying techniques or GPS and have a very high level of relative accuracy (i.e., the measurements of distances between features could be very close to what they are in the real world). At the same time, unless the data are properly tied into a standardized coordinate system, the level of positional accuracy could be quite low. Again, you need to take care that the data development strategy includes processes to correct the positional accuracy of features and documents the methods used.

Finally, the last type of accuracy important in spatial data design is absolute accuracy, which is a combination of positional and relative accuracy. It is a measurement that defines not only the reliability of a feature's position with respect to other features on the same layer, but also with respect to where it is on the face of the earth.

There are fundamentally two different levels of accuracy: map- and surveying-level accuracy. In most cases in GIS design you are dealing with map-level accuracy, which is always lower than surveying-level accuracy. Getting a GIS database to surveying-level accuracy means that all the features in the database, and there could be hundreds of thousands or more, must be located using precise surveying techniques, and the cost would be prohibitive. This is why accuracy is of such concern; to attain high levels of accuracy is exceedingly expensive, but inaccurate data may reduce the utility of the database. As with many things there is a tradeoff between accuracy and cost. The relationship between accuracy and cost is nonlinear, so if you can calculate the cost of developing data with an accuracy of ±1 m and then decide you would like to improve that accuracy to 1/2 m, the total cost will be more than twice the original. Nobody knows the exact parameters of this relationship; we just know it is not a linear progression and that it goes up at least with the square of the required additional accuracy. For example, if original design standards accepted a positional accuracy of ±2.0 ft and managers decided they wanted to double that accuracy to ±1 ft, you can be reasonably certain that the cost of data development will increase at least 4 times, not twice, because the data are in two dimensions.

In the United States, the USGS published map accuracy standards in 1941, revised them in 1947 and they have not changed since. The standards for horizontal accuracy are a function of the map scale (see Table 4.6).

The 1947 map accuracy standards were adequate when most geographic information existed in paper form but not for digital spatial data. In the late 1990s the U.S. federal government developed a new National Standard for Spatial Data Accuracy. For horizontal data the accuracy is based on root mean square error (RMSE). To determine the RMSE for a set of coordinate values you calculate the square of the difference between the coordinate values in the data set and the coordinate values as determined by an independent source of higher accuracy. You sum these squared differences across the data set and take the square root of that sum, which is the RMSE. The accuracy is reported in ground distance (in the data set's distance units) at the 95 percent confidence level. This means that 95 percent of the locations tested to determine accuracy are within that distance of their locations as determined by the higher-accuracy system. In practice, the higher-accuracy system is either GPS or surveying. There is also a reporting procedure for elevation data.

Table 4.6 Table Map Accuracy Standards

Horizontal Accuracy		No More Than 10% of the Points Tested Shall Be in Error of More Than		
Representative Fraction	Feet/inch	Inches on the Map	Feet on the Earth	Meters on the Earth
1:240,000	20,000	0.020	400.00	121.92
1:120,000	10,000	0.020	200.00	60.96
1:24,000	2,000	0.020	40.00	12.19
1:12,000	1,000	0.033	33.33	10.16
1:2,400	200	0.033	6.67	2.03
1:1,200	100	0.033	3.33	1.02
1:600	50	0.033	1.67	0.51

Vertical Accuracy. Regardless of scale, no more than 10% of the elevations tested shall be in error of more than one half the contour interval.

Source: USGS National Mapping Program – National Map Accuracy Standards.
rockyweb.cr.usgs.gov/nmpstds/nmas647.html

Because so much of the source material for GIS databases comes from already published paper maps, often the accuracy you can attain in your GIS is determined by the scale of those maps and whether or not they were prepared using the map accuracy standards for that scale. There should be a statement to that effect on the map; if it is not there, you can make no assumptions about absolute locational accuracy of that data source. But as is clear from the table, if your data source is existing maps, you need very large-scale mapping to attain high accuracy levels. If you extend that table to a ridiculous extent and state that you want centimeter accuracy, you would have to have a map scale of 1:1 (i.e., the map would be the real world). Generally, any desired accuracy below 1 m will require direct surveying of your features.

Precision

After accuracy you need to take into account the precision of the features in your data development strategy. Precision is the ability of a measurement to be reproduced and the number of significant digits to the right of a decimal place a feature can be reliably measured to. This issue of precision is frequently misused in GIS. It is very easy for the user of a system to change configuration settings and have the computer provide locational and distance measurements that can vary from no decimal places to two decimal places to eight places when measuring the same feature. This does not mean that the feature's precision has change, only that the user has altered the way the computer represents the precision. What determines the precision of a feature are the methods used to collect and input the feature into the system. For example, a surveyor or inspector could measure the foundation of a house after it has been built to determine its size. Measurements could be taken to the one-hundredth of a foot and recorded in a field book. When this person returns to the office and draws the foundation into the system, it can be drawn using coordinate geometry (COGO). If properly measured and recorded in the field, the feature would close and create a polygon representing the foundation when the measurements were checked. This would indicate a high level of relative accuracy. When in the field, the worker could also record three (or more) GPS points to determine the foundation's location in a standard coordinate system. These points could then be used to locate the foundation in the GIS.

The precision of this foundation and any of its sides would be one-hundredth of a foot, but the precision of any measurements taken from this foundation to any other feature measured and located in the same manner would be based on the precision of the GPS locations take to locate the corners. If these locations were accurate to 1 m, the precision of measurements between two foundations would be ±2 meters. But if the corners were located to within one-hundredth of an inch, the measurements between foundations would also be that precise.

Completeness

The third related issue is the need for completeness within a data layer. You should design standards for absolute accuracy and precision but also for completeness.

Completeness is defined as locating all the features represented in the data layer or some stated percentage of those features. Locational accuracy standards are stated as the percentage of visible features that are within a certain distance of their real location, and completeness is a statement of the percentage of the actual features that are captured in the database. It is important to state exactly what is the basis for measuring completeness. You would certainly expect to capture all land parcels delimited on the parcel maps for a community as of a certain date, but can you be certain that those maps contain depictions of all the land parcels in the community? There is a probably apocryphal tale told in GIS of a community that paid for its GIS by discovering many land parcels that had not been correctly entered into the taxation database or were missing entirely. The back taxes collected on those parcels were said to have covered a significant percentage of the GIS development cost. It would be nice to be able to document this story, but it probably is not true.

Specifying the desired level of completeness is particularly important when data are to be captured from aerial photography. It is common for features to be obscured by shadow and branch overhang, and although you might wish to collect every catch basin and utility pole, that may not be possible at the scale you have selected. Measuring the delivered positional accuracy and completeness of delivered data are quality control tasks that you need to design in as you are specifying your spatial data. Determining that a set of digital data actually meets a certain map accuracy standard or that you actually have captured 95 percent of the features you expected to capture is a task that is done after completing the data by field work with higher accuracy measurement systems and careful field checking of the presence or absence of visible features. In the design phase you specify your desired levels of accuracy, precision, and completeness, and in the implementation phase you document whether or not you have met these standards.

Accuracy Concerns – Global Positioning Systems

GPS is another very common method for obtaining spatial data to include in a GIS. So it is important in the design phase to specify the standards you expect for data derived from GPS. GPS technology was originally developed by the U.S. Department of Defense and allows the user to determine exactly where they are on the face of the earth by triangulating data from a series of satellites that orbit the earth. Although the military applications drove the construction of the system, civilian use far outstrips military use of GPS around the world today. The user has a GPS receiver that receives the signals from a constellation of these satellites, processes them, and then calculates the latitude and longitude of the receiver at the time the data were collected. The accuracy of the data that can be acquired from this process depends on the type of receiver, the methods that are used to collect the data, the amount of time that data are collected, and what, if any, post-processing is done on the data with more advanced software. If you are trying to get meter-level accuracy, there are eight major areas in the system that can generate error; getting more accurate than that, the list gets even longer. Error in GPS is a complex topic, and we only introduce it here in general terms.

The primary components of the system are the satellites, ground stations, receivers, and software used to correct known errors in the data. The satellites make up a worldwide network of 24 satellites that orbit the earth every 12 hours. With this number of satellites moving at that speed there are almost always enough satellites in direct line of sight of a receiver that determining location is possible. The signal broadcast by the satellites contains information about exactly when the signal was sent, and information that tells the receiver which satellite it is listening to. By comparing the time it receives the signal to its internal clock, the receiver can calculate its distance to the satellite.

If the earth were not nearly spherical, it would require this distance information from a minimum of three satellites to determine location in latitude and longitude. Let's say we measure a distance of 10,000 km from the first satellite (the satellites actually orbit at about 11,000 nautical miles, so these numbers are just for illustration). This first distance defines a sphere with a radius of 10,000,000 km and the receiver must be on the surface of the sphere. Now, the distance from the second satellite, let's say 12,000 km, locates the receiver on the surface of a second sphere with a radius of 12,000 km. The intersection of these two spheres defines a series of locations where the sphere must be. Calculate the distance to the third satellite and, if the clocks were perfectly in synchronization, the number of possible locations is narrowed to two points where the receiver may be. One of those points is on the face of the earth (assuming that is where the receiver is), and the other is usually some where in space.

To resolve this conflict and because it is not possible to have clocks telling exactly the same time on both the satellites and the receivers, a fourth satellite is required to eliminate the incorrect location and correct for the clock differential. Because it takes between .05 and .07 seconds for the signal to travel the 11,000 nautical miles, accurate clocks are critical, and the clocks on the satellites are large and accurate atomic clocks. Because you cannot (financially or physically) place an atomic clock on a receiver, a mathematical correction is made in the receiver. If the clocks were perfectly synchronized, the four spheres would intersect at a single point, the location of the receiver. Because of clock error they intersect in a three-dimensional surface, and the receiver applies a correction to its clock until the points all intersect at a single location, which it reports to the user. All this happens almost instantly at the push of a button after the receiver has locked on to at least four satellites. This makes GPS receivers very accurate clocks as well.

The United States military, which controls the GPS satellite system, can degrade the signal so that only military receivers will get accurate locations. This is known as selective availability, and when it is on, the signal is degraded. Since May 2000 it is now always turned off, so there is no difference between positional accuracy from military and civilian receivers. The chief of the U.S. Geodetic Survey has related this to a football field; if selective availability is off, you can only know that you are on or off the field, but if it is on, you can tell at which yard marker you are located.

Another important factor to making GPS work is knowing the exact location of the satellites at all times. For low-end or hand-held GPS receivers such as those used by hikers and boaters this is not important, but for survey-grade receivers, those typically used to collect GIS data, this is important. The U.S. Department of Defense monitors these satellites very exactly using radar and records the position, altitude, and speed of the satellites. Along with the pseudo-code mentioned above, the satellite also broadcasts this information, which is known as the ephemeris errors. The errors themselves are usually pretty small, but if you want exact measurements, they need to be corrected for.

All of this discussion is based on the premise that you can easily calculate the distance from a receiver to a satellite. The problem is that the signal itself has to travel from the satellite, through the atmosphere, and a number of factors can affect the speed of the signal, including the temperature of the atmosphere and how much moisture is in it. Now, if all of these factors could be known, they could easily be accounted for with mathematics, but the reality is that these factors are difficult if not impossible to track. The way these are accounted for is by tracking two different signals from the same satellite and calculating their relative speed. For this reason, the receivers that are used to perform accurate measurements need to be dual frequency where the information is carried twice on different frequencies. A dual-frequency receiver can measure two signals from the same satellite, know the time that each signal left the satellite, and calculate correction factors to apply to the results to reduce error in measurements. These types of receivers cost more but provide much improved accuracy in measurements.

Ground stations are permanent locations, also known as the control segment, that are in constant communication with the satellites. These control stations transmit signals to the satellites to check their exact position in space and assure that they are operating properly. The signals sent to them are then used by the satellites to adjust the position and the signals that they send out to the GPS receivers on the surface of the earth. Five ground stations exist and are located in Hawaii, Ascension Island, Diego Garcia, Kwajalein, and Colorado Springs.

Differential Processing

Another type of ground station, much more widely distributed, is known as a base station. These are locations with highly accurate and expensive GPS receivers that constantly monitor their location as determined by the satellite. Because of all the possible errors in the system, the location changes slightly depending on atmospheric conditions, satellite errors, and other factors. But the location of the base station is know by surveying methods to a very high degree of accuracy, so it is possible to mathematically model the deviations from the known location compared to the location estimated at the same time your receiver was taking measurements. Of course, you want to select a base station near to where you are taking your measurements, and there is a widespread and growing network of base stations around the world. Both receivers, the base station and your receiver, are communicating with the satellite at the same time. and this processing is known as differential GPS (DGPS). This processing can occur either in real time by having FM radio links between your receiver and the base station or on a computer after you have taken your measurements, known as postprocessing DGPS. Many federal and state agencies have established permanent base stations that transmit their locational data constantly and store the files on servers tied to the World Wide Web for postprocessing. Using DGPS and postprocessing you can anticipate measurement results within a couple of meters for moving applications where few points are taken at each location. The results can be within subcentimeters if you use dual-frequency real-time DGPS and stay at one location for an extended time. GPS has not yet evolved to the point where you can stand on top of a feature and push a button and in one or two seconds receive locational information accurate enough to locate that feature within a centimeter of where it is, but the systems get closer to that goal all the time.

Real-Time Kinematic GPS

Real-time kinematic (RTK) GPS is an advanced type of differential GPS. With these units a combination of a stationary base station located at a known coordinate location and a roving RTK receiver, usually in a backpack, are used at the location where you want to calculate new coordinates. The difference between a DGPS and an RTK unit is that the RTK roving unit communicates back to the stationary unit using radio waves, and the unit can calculate its location much faster and more precisely. These type of receivers are typically much more expensive, but because of their efficiency, they are often used to collect data for a GIS where a large number of points need to be collected; the positional accuracy is very high (less than a meter). A typical application of this type of RTK GPS is utility infrastructure mapping such as water valves or utility manholes and catch basins.

With hand-held units and no differential processing you can achieve accuracies between 3 to 5 m, or 10 to15 ft, which approaches map accuracy values at the 1-inch-to-200-ft scale. Differential processing and hand-held units that allow you to process the data this way can now achieve accuracies in the 2- to 5-meter (6- to15-ft) range. Submeter accuracy using RTK GPS is very expensive and requires sophisticated equipment, real-time information from multiple base stations, dual-frequency radio, and complex mathematical processing, but submeter accuracy is getting more affordable over time.

Accuracy across Layers

The largest concern relating to accuracy, however, is not determining how accurate one layer of a set of features may be but how to assess accuracy when there are multiple layers in your GIS, each with differing levels of accuracy. This has been stumping the profession for years. What is commonly done is to take the most accurate data and use that data source to realign the inaccurate information. For example, often in GISs that use aerial photography as a backdrop or data source, that is the most accurate information in the database. Features from other sources can be shifted to fit visible features on the digital photograph (see Chapter 6). This process of conflating one data set to fit another improves the visual fit of data layers, but you still cannot state that the layer is necessarily more accurate than it was before. It probably is, but all you can say about it is that you performed that function on the data to improve the fit.

In the end, the level of accuracy you require is determined by the uses to which you plan to put the GIS. For many uses (e.g., regional planning over multiple states), available data at a scale of 1:24,000 (accuracy ±40 ft) is adequate. For city planning in dense areas you may need data accurate to a map scale of 1:1,200 (±3.3 ft). The issue of accuracy often comes down to differences of opinion between engineers, who have been trained to strive for the highest accuracy possible, and others who would be satisfied with the lowest level of accuracy that keeps them from making bad locational decisions. How those tensions get resolved and the resources you have to invest will determine the base accuracy you will be seeking in your database. Keep in mind that most GISs are not built to be design tools. Rather they are intended to be a model of the real world. The real world is a messy place, and sometimes a high level of locational accuracy is not only unnecessary but also unattainable.

Choosing a Coordinate System and Map Projection

A human being, as an object, has the properties of height and weight, and we can measure those properties in different measurement systems such as inches, meters, pounds, and kilograms. Similarly, geographic objects have the property of location, and we can measure that property in different coordinate systems and map projections. The correct choices of coordinate system and projection are important because this is how you will present your data to the users and how you will take measurements from the data. It may seem that the question "How many counties in the United States are within 100 miles of any part of Franklin County, Ohio?" has a unique answer, but it does not. If you begin with source data that came from maps at a scale of 1:2,000,000 and use an Albers Equal Area conic projection with the Clark 1866 spheroid, a central meridian of W96, a reference latitude of N 40, and standard parallels of N 20 and N60, you get an answer of 122 counties, including Franklin. This projection is designed to show the United States well in the larger context of North America so that Alaska will show. But if you project the data with the same type of projection but with reference latitude of N 37.5 and standard parallels of N 29.5 and N 45.5, a projection designed to display the conterminous United States, you get an answer of 116. And if you use the geographic projection (i.e., latitude and longitude), you get an answer of only 90 counties. How many counties are within 100 miles of Franklin County, Ohio? Well, the answer depends on the projection used for the data. A basic knowledge of map projections is essential for all map users and makers, "no matter how much computers seem to have automated the operations," and it holds even more for GIS database designers (Snyder 1987, 3). Although there are many terms specific to map projections, Table 4.7 is a simple glossary of the terms use in this section.

Table 4.7 Some Basic Projection Terminology

Developable surface	A three-dimensional feature that can be broken apart and which will lie flat; a sphere is not one, whereas a cylinder is.
Conformal	Property of a map projection such that relative local angles around every point on the map are shown correctly.
Equal-area	An area on the map covers the same area on the earth no matter at all locations on the map.
Geoid	The shape that the earth would take if it were all measured at mean sea level. Newer projections use a mathematical representation of the geoid rather than any particular spheroid.
Ellipsoid or spheroid	The earth is nearly spherical but not quite; it is elliptical and slightly flattened at the poles. There are over 25 spheroids with different semimajor and semiminor axes used in different parts of the world and at different times. Ground-measured spheroids are still in wide use for specific areas although there are several satellite-measured spheroids designed for use around the entire globe.
Semimajor axis	Equatorial radius.
Semiminor axis	Polar radius; always smaller because of flattening.
Flattening	Difference between the two axes expressed as a percentage of the semimajor axis.
Datum	A set of parameters that define a coordinate system. A spheroid is part of a datum, but many datums could be based on single spheroid. Local datums, such as the North American Datum of 1927 (NAD27), have a central point, and all locations are measured relative to that point. Local datums are aligned to the spheroid to fit locations in a particular area. A geocentric datum such as the World Geodetic System of 1984 (WGS84) uses the center of the earth's mass as the reference point.
NAD27	A local datum based on the Clarke 1866 ellipsoid; defined for used in North America with a reference point at Meade's Ranch in Kansas. It is being phased out and replaced with the North American Datum of 1983 (NAD83).
NAD83	A geocentric datum, both satellite and ground-based, using an ellipsoid developed from the Geodetic Reference System of 1980 (GRS80). Gradually replacing NAD27 as the standard datum in North America.

continued

Table 4.7 (Continued)	
UTM	Universal Transverse Mercator. A set of transverse mercator projections adopted by the U.S. Army in 1947 for global large-scale military maps. Between N 84 degrees and S 80 degrees the globe is divided into 6-degree segments numbered 1 to 60 and 8-degree latitudinal segments using letters. The measurement units are meters; the central meridian is the midpoint meridian between the bounding meridians and is given a false easting of 500,000 m. N/S coordinates are measured from the equator with a y value of 0 for the northern hemisphere and 10,000,000 for the southern. Numbers increase going east and north, avoiding negative numbers.
Standard parallel	In conic projections, the one or two parallels (lines of latitude) along which scale is true; at latitudes off the standard parallels map scale varies.
Central meridian	The line of longitude in the middle of the projection; x locations are measured from this line.
Latitude of origin	Line of latitude representing the 0 or base line for measuring distance in the y dimension.
False easting	Arbitrary number given to the central meridian to avoid negative numbers
False northing	Arbitrary number given to the latitude of origin to avoid negative numbers.

Decimal Longitude and Latitude or Projected Data

The first decision is whether to store your data unprojected, in decimal latitude and longitude, and project on the fly as you display the data or to store your data in a projection and coordinate system. As your area of interest gets larger, incorporating more of the earth's surface, choosing to store your data in longitude and latitude is more reasonable. Traditional use of latitude and longitude describes locations using degrees, minutes, and seconds with a leading text value to distinguish between north or south latitude and east or west longitude. Latitude is usually presented first. The convention now is to use decimal notation rather than minutes and seconds, with negative values for south latitude and west longitude. So a location in the Brazilian Amazon that might be recorded in a gazetteer as S 7 degrees 50 minutes 20 seconds, W 63 degrees 30 minutes 15 seconds in decimal notation will be −63.501467, −7.838889. Notice also that longitude is given first because it represents the x dimension in a plane coordinate system, and latitude, representing the y dimension, is reported second; this is backward from traditional gazetteer notation.

A common reason to store your data in decimal longitude and latitude is that your area of interest is extremely large, covering many states or provinces or even continental in size. Because you may be working at many different spatial scales and in many parts of this large area, you will want to display and analyze the information in a projection appropriate for the subregion in which you are working. If your GIS was designed to cover all of North America and you chose to use the standard Albers Equal Area projection for North America and were working with data showing Washington, state, you would find the entire state appears tipped up to the northwest about 15 degrees. If you project the data to the state plane coordinate system for northern Washington, the state appears to run more directly east to west, which it actually does. Because the border between Washington and British Columbia is a line of latitude for much of its distance, this is what people expect to see. Projections on the fly from longitude and latitude are available in most GIS software systems, but projection transformations from a plane coordinate system to another plane coordinate systems require explicit processing and duplication of the data. If you have data in a projected coordinate system, you cannot display it in latitude and longitude on the fly.

Storing data in latitude and longitude is also a sensible decision if your data are coming from many different sources, each of which may provide data in a different projection

If you transform the data to latitude and longitude before inserting it into your database, the data will all fit together, but if you put it in without transforming it to latitude and longitude (or another common projection), one set of data will appear to draw on top of the other or otherwise be out of location. Figure 4.10a shows the statewide road files from Connecticut and Rhode Island as they are stored and used by the respective state GIS agencies using two different projections. Because each uses its own state plane projection system, the two sets of data do not appear in correct relative location. After transformation of both to latitude and longitude, they appear correctly located relative to each other (see Figure 4.10b).

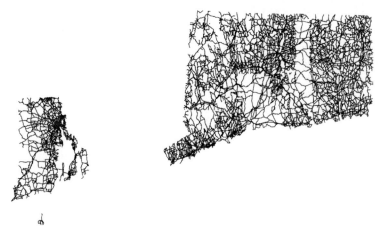

Figure 4.10a Connecticut and Rhode Island major roads – state plane projetions.

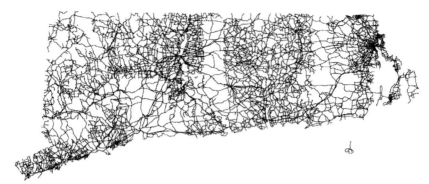

Figure 4.10b   Connecticut and Rhode Island major roads – latitude and longitude.

If you choose to store your data in geographic coordinates and allow the software to project them as you display them, the next decision will be the storage precision you will need. Computers store floating-point data (numbers with decimal points) in either single or double precision. In single-precision storage there are seven significant digits; this allows you to store data that are accurate to around one-half an inch. So single-precision data storage, which requires fewer resources to store and process, is almost always adequate for spatial data represented in latitude and longitude. Many believe that double-precision storage is better than single-precision storage, and for some scientific applications that deal with extremely large or small numbers this is the case. But in GIS there are very few applications where the accuracy of the data requires any use of double-precision numbers.

Characteristics of Map Projections

Some practitioners believe that all spatial data should be stored in decimal latitude and longitude because it facilitates transfer and can be converted to any other projection using standard transformation tools. Most, however, store their data in a particular projection and in measurement units such as feet or meters. Map projections exist because the Earth is essentially a sphere, a three-dimensional object, and a paper map or computer screen is a two-dimensional object. Some three-dimension objects, such as a cube, will unfold and lie perfectly flat. Three-dimensional objects (cones and cylinders) that can do this are known as *developable surfaces*. A sphere or spheroid is not one of these surfaces, so map projections are the mathematical transformation of the sphere to a flat surface. Distortions or compromises of various sorts occur because the sphere will not lie perfectly flat. Because this cannot be done without distortion, the choice of map projection is fundamentally a choice of which distortions your applications can survive. For large-scale work (small areas) the choice of map projection is not as important. At a scale of 1:24,000 on a paper map more distortion is introduced by expansion and contraction of the paper due to humidity than by choice of map projection (Snyder 1987). For small-scale work (large areas) the choice of projection becomes more important. The characteristics you need to consider when choosing a projection are as follows:

- *Area.* Some projections are *equal-area,* which means that an area of a certain size on the map covers the same amount of area on the earth's surface no matter where on the map it is. This forces some distortion in the other map properties.

- *Shape.* Some projections are *conformal,* which means that relative local angles around all points on the map are correct. On conformal maps lines of latitude and longitude cross each other at right angles, but areas are enlarged or reduced except along certain lines in some projections. All the large-scale maps used by the USGS and most other agencies of the U.S. federal government involved in mapping or GIS are conformal projections, and all state plane coordinate projections in the United States are conformal. No map or set of geographic data can be both equal area and conformal.

- *Scale.* Map projections do not show scale correctly throughout the map, which is why you will get different measurements between the same two points depending on the projection used. There are usually one or two lines along which scale is correct, and if you choose those correctly, you can minimize the scale variation across the data or the map. There is a special class of projections, *equidistant,* where scale is correct to all points on the map from one or two specified points on the map or along all lines of longitude.

- *Direction.* On a group of map projections known as *azimuthal* the directions of all points on the map are correct with respect to one location at the center of the map. On other map projections direction is distorted.

Spanning Existing Map Projection Zones

Most large-scale (small area) applications select conformal projections and, in the United States, many organizations will choose from the set of state plane coordinate systems (SPCSs). For the smaller states and a few of the larger ones there is one projection. Problems arise when your area of interest spans two or more defined projections as in the example with Connecticut and Rhode Island demonstrate (Figure 4.10a and b). This situation occurs within states and provinces; it is quite likely, for instance, for a municipality to have part of its territory in one state plane coordinate zone and part in another. This is even more likely when the area of interest includes multiple units of government.

If you are in a mid- or low-latitude region, there are relatively few solutions to this problem. Locations at high latitudes require different projections. If only a small part of the area is in one projection region, you can use the projection of most of your area of interest. This would entail transforming the data from the smaller projection area into the projection of the larger. But if your area of interest is rather evenly split between two projections, you will have to create your own projection.

Follow the procedure in Table 4.8. This will produce a projection that will allow data from either of the input projections to display in correct location relative to each other. If your area covers more than two projection areas, the decisions are a little more complex but similar. Again, this process is appropriate for small areas in low or midatitudes.

Selection of Projection for Large Areas

It is not easy to specify when small becomes large and choice of map projection becomes more significant. Generally, over small areas of 6 or 7 miles across, changing from one projection to another is hardly noticeable, certainly not visually, and if your data source is paper maps, physical distortion in the paper is greater than projection variation. At this scale differences in projections result in distance differences of only up to 2 ft per mile. When the area of interest extends to a hundred miles or more, measurements of distance can vary significantly as we saw in the example of finding counties with 100 miles of Franklin County, Ohio. For regions that include an entire nation of the world, there are usually preferred projections that have been chosen by the principal mapping agencies of the nation. Areas smaller than a nation but larger than provinces or states usually do not have preselected projections. The decisions are similar to those in Table 4.8, but there is usually no opportunity to use parameters of existing projections such as false eastings or northing, latitudes of origin, standard parallels, or central meridians. This means the designer must choose these parameters of the projection. For a transverse mercator projection when extent of the area of interest is principally north/south, the parameters are as follows:

- *Central meridian.* Select a line of longitude (it may include minutes and seconds expressed as decimal degrees).

- *Latitude of origin.* Should be a parallel to the south of the area of interest. It can be the equator, as it is in the Universal Transverse Mercator (UTM) family of projections, but this results in very high y values in the millions of meters or tens of millions of feet. This set of projections was adopted by the U.S. Army in 1947 and covers the world between latitudes 84 degrees north and 80 degrees south.

- *Scale factor at central meridian.* Selecting a value of 1 means that scale is accurate only along this line. A value of less than 1 means that scale is accurate along a pair of lines on either side of the central meridian. The UTM projection uses 0.9996 as the scale factor.

- *False eastings and northings.* Select values in the chosen measurement units, usually feet or meters, to avoid negative numbers. The false easting value is assigned to the central meridian as its x value and the false northing to the latitude or origin as its y value; x and y values decrease and increase from these origins.

Table 4.8 Projection Selection. Is the Principal Extent of the Area of Interest North/South or East/West?

North/South	East/West
Select a **transverse mercator projection.** These projections have the following properties:	Select **lambert conformal conic.** This projection has the following properties:

North/South column:

Select a **transverse mercator projection.** These projections have the following properties:

Small shapes are maintained but larger regions are distorted away from the central meridian. Areal distortion increases with distance from the central meridian. Local angles are accurate everywhere. Scale is accurate along the central meridian if the scale factor is 1.0. If the it is less than 1.0, then there are two straight lines having an accurate scale, equidistant from and on each side of the central meridian.

Selection of Parameters:

Datum: Use the datum for the original projections

Spheroid: Usually included as part of the datum if appropriate. Small area projections always use a spheroid or the geoid.

Measurement Units: Feet or meters; preferable to use the same as the original projections

Central Meridian (line of longitude): Select the nearest whole degree of longitude in the center of the area of interest. This will be different from either of the original projections.

Latitude of Origin: Use the same value as for the original projections unless your area of interest extends to a lower latitude. Then use a different value. It is common to use whole degrees or quarters of degrees (e.g., N 41 degrees 45 seconds).

Scale Factor: See properties above. For small areas it should be 1.0

False Easting: Use the false easting of the original projections unless this will result in negative y (longitude) numbers. In that case the number will have to be increased. All false eastings (and northings) use round numbers (e.g., 500,000).

False Northing: Use the same value as original projections.

East/West column:

Select **lambert conformal conic.** This projection has the following properties:

All intersections of latitude and longitude lines are 90 degrees. Small shapes are maintained.

Minimal distortion of area near the standard parallels. Scale is correct along the standard parallels but smaller between then and larger beyond them. Local angles are accurate.

Selection of Parameters:

Datum: Same as North/South

Spheroid: Same as North/South

Measurement Units: Same as North/South

Central Meridian (line of longitude): Same as North/South

Latitude of Origin: Same as North/South

Standard Parallels: If your area of interest has the same North/South extent as both the original projections, you can use the standard parallels of those projections. Otherwise choose your own standard parallels. The general recommendation is one-sixth of the length of longitude of the area of interest from the northern and southern ends.

False Easting: Same as North/South.

False Northing: Same as North/South.

For a Lambert conformal conic projection when the extent of the area of interest is principally east/west, the parameters are as follows:

- The concerns for the central meridian, latitude of origin, and false eastings and northings are the same as for the transverse mercator.

- Standard parallels should be selected one-sixth and five-sixths of the distance from the southern edge of the area of interest. Scale is correct along these two parallels and varies with distance from them.

The issues around projections are important, and this concern for projecting the real world onto a plane projection is one of the things that make spatial data different. In a database, spatial data are not so special; it is just another form of data. But the design considerations for projecting this data are special, and a basic knowledge of map projections is important. A few key points in summary are as follows:

- If your area of interest is entirely within a single predefined, governmentally approved projection, it would make sense to select that projection unless there are reasons not to do so. Local governments using the SPCS for their area is an example.

- Some organizations and projects require data to be submitted in a particular projection, and if your GIS is part of this project, you should use that projection. For instance, many scientific projects in the United States store data in UTM projections with measurement units in meters.

- If your area spans two governmentally defined projections but most is in one of them, select that one. If they are evenly split construct a similar projection.

- If your area of interest is a high latitude (above 80 degrees north or south) neither the Lambert Conformal Conic nor the Transverse Mercator projection is appropriate and you should consult some of the sources listed.

Whatever map projection and measurement units you select, it will be almost certain that you will be taking in digital data in other projections, so you almost always need to have the capability of transforming data from one projection to another. Some of the less expensive GISs do not have this capability, but there is relatively inexpensive software available that incorporates the standard, public-domain transformation equations. Choosing the wrong map projection is a serious mistake; it is possible to transform from the old projection to a better one, but if you have much data, this is a bothersome process.

Although the mathematics and choices around map projections seem very complex and the number of possible projections very large, if your area of interest is small, there is usually a small set of choices to make. Your reason for choosing a projection will almost always be to preserve shape or area (conformal or equivalent projections) and rarely to maintain straight line of movements for correct scale

Table 4.9 Selection of Projections – Small Areas

Projection Class		
Orientation of Area of Interest	Conformal (Preserves Shape)	Equal-Area (Preserves Area)
Predominantly east–west (away from the equator)	If you are inside the zone of a single existing projection, use it. It will probably be a Lambert Conformal Conic projection with parameters to fit the existing zone (e.g., State Plane). Outside an existing zone or between two or more, create a Lambert Conformal Conic projection with two standard parallels one-sixth of the length of longitude of the area of interest from the northern and southern ends.	Same process but use an Albers Equal Area projection.
Predominantly north–south (away from the equator)	An existing projection in a single zone will probably be a transverse mercator or UTM. If you span zones, create a transverse mercator projection using the central meridian of your area of interest, a latitude of origin below your area of interest, and a false easting large enough to ensure no negative numbers west of the central meridian.	Mercator and transverse mercator projections are conformal, not equal-area.
Predominantly oblique (some direction other than EW or NS)	Oblique mercator projection.	
	Whereas a standard mercator projection is "wrapped around" the equator, and a transverse mercator projection is wrapped around the poles, an oblique mercator projection is wrapped at an angle. See a projection manual.	Mercator and transverse mercator projections are conformal, not equal-area

Source: Based on information in Snyder 1987.

along meridians or lines or latitude or straight great-circle routes. You merely want a projection that will fit your portion of the earth well and produce planar maps and computer displays. And most, though possibly not all, of the readers of this book will have an area of interest away from the equator. When you restrict projection choices by that much, it becomes easier. Table 4.9 should serve as a basic guide.

Spatial Indexing

Another design concern for spatial data is indexing of the data. Indexing allows rapid retrieval of features from large data sets, and without it a GIS may operate so slowly that users will not use the system. GISs automatically index the spatial information, so it is really not necessary to understand it unless your database is so large that the default indexing scheme is not adequate or super-fast access is critical to your applications. In those cases you need to set up the spatial indexes with considerable care. Efficient spatial indexing has been an area of research and development in computer science for almost 30 years, and new techniques for indexing spatial data are regularly proposed. Until they get worked into the existing GISs and RBDMSs, though, they remain essentially research tools. The spatial indexing process discussed here is typical of those implemented currently in commercially available GISs.

As with an attribute-indexed query, (see chapter 5 – Principles for Fields in D Tables) a spatially indexed query for features reads from a structured index table or tables and then, through relational joins between the index tables and the feature table, pulls out only the features that have been requested. Almost every interaction with the data in the map view is a spatial query. Panning the map from one location to another is the equivalent of

SELECT ALL FEATURES FROM THESE_LAYERS AT THIS_SCALE WITHIN A RECTANGLE OF THIS_X_DIMENSION AND THIS_Y_DIMENSION CENTERED ON THIS X_Y_LOCATION.

Every time you pan the map view, that query is sent again to the database with a different value for the center of the view. Zooming in or out is a spatial query as is selecting features inside a circle drawn on the map view. All these queries require reads against the feature database, which is not ordered geographically, and by reading against the indexed tables first, you are able to reach directly into the feature table to retrieve the requested geographic features.

The simplest kind of spatial index is created with a set of tiles that exclusively and exhaustively cover the area of the geographic features. *Exclusively* means that all parts of the region of interest are covered by only one tile and *exhaustively* means that no area of the region of interest is uncovered by a tile. This is called a tessellation. In Figure 4.11 the tiles are numbered T1 through T12.

The spatial query being executed here is a so-called window query that seeks to find and display all the features inside the window you have drawn on the screen. The search determines which tile(s) touch the query window (in this case, T1, T2, T5, and T6). It then goes into the index table to find out which geographic features are inside these tiles and retrieves only those features from the data set (i.e., features A, B and C). All these features are retrieved from the T1 record in the index table; the other records, T2, T5, and T6, would have to be read, but no new features would be added to the already selected set. Because tile number organizes the index table, that first search of the index table is actually an indexed search on the tile number. With no index the entire feature table would have to be read and each feature evaluated to see whether it is inside the query window or not. There are only five features in this simple example, but what if there were several hundred thousand and you were interested only in a small area? One feature retrieved, A, is actually not inside the query window but would be retrieved anyway because it, or at least part of it, is inside tile 1. Slightly more sophisticated indexing systems will execute a second query at this time of selecting from this set (A, B, C) only those features that intersect the query window. This will result in A being thrown away and not displayed and B and C being the only features returned. One GIS system, ArcInfo, uses a three-step indexed search where each feature is a record in a feature envelope table that contains the x/x pairs of the corner of a rectangle that completely encloses the feature. The first stage of the search pulls the feature envelopes that touch the selected tiles; the second stage eliminates those feature envelopes on one of the selected tiles but outside the query window; the third stage compares the actual features that remain to identify those that are inside the query window.

Because the simplest indexing scheme would have retrieved a feature that is not in the requested set, it is clear that choice of the size of the tile is important. In most cases the default tile size will result in adequate results. If you make the tiles too large, as in this case, you retrieve excess features, but if you make them too small, you create a much larger index table; searching that table slows retrieval down. More complex indexing systems support three levels of tile sizes where there are three index tables and the one searched is determined by how far out you are zoomed in the display. So if you draw a large query window with the entire area of interest in view, the largest tile is searched, but if you draw a small window, the search accesses the index table based on the smallest of the tiles. If you have literally hundreds of thousands of features to search and typically are zoomed into a small set of x/x ranges (zoom levels), you could set up the tiling system to approximate the extent of the areas displayed at these zoom levels as your three levels of tiling.

Setting up the spatial indexes and correctly locating the various files that need access for queries is referred to as *tuning* the database. Tuning is a craft that database professionals excel at, and the addition of a spatial index to already existing indexing schemes is something they will quickly understand and implement. Locating the files that are read frequently in the center of the storage medium (disk) and those that are read less frequently on the periphery will also speed access, as will partitioning or splitting very large feature tables onto several different disk storage devices, so long as each with its own disk controlling hardware and software.

Often the size of the database is so large that access is not determined as much by the speed of the CPU or the amount of RAM you have as by the physical/mechanical limitation of disk access. The setup and tuning of very large databases (VLDBs) is a specialized craft and beyond the scope of this book, so seek professional help when the number of features you need to query regularly and quickly extends into the hundreds of thousands. Multitile indexing, careful file placement, and partitioning of large tables onto several disks are necessary for rapid retrieval in that situation.

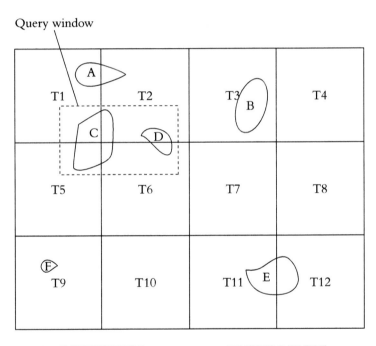

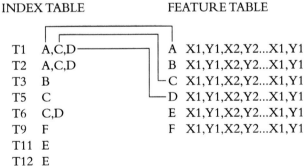

Figure 4.11   Simple spatial index.

Conclusions

It is because GISs must deal with spatial data that it is a distinct area of inquiry in database design and implementation. It is difficult and complex to deal with spatial data in flat tables and object-oriented databases. Users are so used to the bird's eye gestalt view we have had of geographic data (the map) that we tend to underestimate what it takes to store that information in tables and be able to extract it into a display. How you design your spatial data determines how useful your GIS will be. Some design mistakes are relatively easy to correct, but the design of your spatial data is so basic and central to the system that it is difficult to alter it after construction. It is like a foundation or skeleton for the entire database. For example, initially constructing a GIS database using a set of layers and then changing your mind and deciding to build a SERD out of these layers is almost a complete redesign. At a minimum you should have answered these questions:

- Are we going to build this database out of layers, or are we going to construct a SERD?

- What is the geographic extent of the area we need to cover in this database (i.e., which portion of the real world are we going to model)?

- What is the appropriate mix of data models, raster and vector, that will best support our applications?

- What are the best projection and measurement units for our data set?

- What type of spatial feature we will use to represent each layer or set of real-world features? Do we need any special features of representation such as multiple point, line or polygon features, planar networks, or dynamic segmentation?

- How accurate are the various layers and feature sets in our database going to be?

More questions will arise during the process, but if you can provide answers for those questions, you are a long way to an appropriate design for your spatial data.

ADDITIONAL READING

Burrough, P. A., R. MacDonnell, and P. A. Burrough. 1998. *Principles of Geographical Information Systems,* 2nd ed.. Oxford University Press: Oxford, UK.

De Floriani, L., B. Falcidieno, G. Nagy, and C. Pienovi. 1984. A hierarchical structure for surface approximation. *Computing & Graphics* 8(2):183–183.

Heller, M. 1990. Triangulation algorithms for adaptive terrain modeling. In *Proceedings of the 4th International Symposium on Spatial Data Handling*, 163–174.

Field, D. A., and W. D. Smith. 1991. Graded tetrahedral finite element meshes. *International Journal for Numerical Methods in Engineering,* 31(3):413–425.

Goodchild, M. F., and K. K. Kemp, eds. 1990. NCGIA Core Curriculum in GIS.

National Center for Geographic Information and Analysis, University of California: Santa Barbara, CA.

Laurini, R., and D. Thompson. 1992. *Fundamentals of Spatial Information Systems.* Academic Press: San Diego, CA.

Lowell, K., and A. Jaton, eds. 1999. *Spatial Accuracy Assessment: Land Information Uncertainty in Natural Resources.* Ann Arbor Press: Chelsea, MI.

Maling, D. H. 1973. *Coordinate Systems and Map Projections.* London: George Philip & Son, Ltd.

McHarg, I. 1969 *Design with Nature.* Natural History Press: Garden City, NY.

Peuquet, D.J. 1990. A conceptual framework and comparison of spatial data models. In *Introductory Readings in Geographic Information Systems.* Bristol, PA: Taylor & Francis.

Robinson, A.H. et al.1995. *Elements of Cartography,* 6th ed. John Wiley & Sons: New York.

Samet, H. 1990. *The Design and Analysis of Spatial Data Structures.* Addison-Wesley: Reading, MA.

Snyder, John P. 1987. *Map Projections — A Working Manual.* U.S. Geological Survey Professional Paper 1395. Washington, D.C.: U.S. Government Printing Office

Stem, J. E.1989. *State Plane Coordinate System of 1983.* NOAA Manual NOS NGS 5. Washington, D.C.: National Oceanographic and Atmospheric Administration.

INTERNET RESOURCES

Spatial data transfer standards components:
ftp://sdts.er.usgs.gov/pub/sdts/standard/latest_draft/pdf/

Global positioning frequently asked questions:
vancouver-webpages.com/peter/gpsfaq.txt

Map projections:
gisdatadepot.com/helpdesk/projections.html

Datums and geodesy:
environment.sa.gov.au/mapland/sicom/sicom/tp_scs.html

Map projections at high latitudes (antarctica):
aadc.aad.gov.au/mapping/map_projections.asp

Projections and coordinate systems, GIS lounge:
gislounge.com/ll/projections.shtml

National Standard for Spatial Data Accuracy:
http://www.fgdc.gov/standards/documents/standards/accuracy/

Design Issues for Attribute Data

The move toward incorporating all data in SERDs notwithstanding (see Design for Spatial Data section on the two models), it is still sensible in GIS to refer to two types of data, spatial and attribute data. The spatial data are the where and the attribute data are the what (see Figure 5.1).

In the layer data model the two sets of data are joined on the unique ID value, and there is a one-to-one correspondence between a geographic feature and the row in a data table with the attribute data. This method of having two separate tables joined on a unique ID field (or some more complex version using more tables) was the standard for many years in GIS software. The principal reason for this was that the software companies stored the spatial data in proprietary data structures, which they did not wish to share with others. The attribute data were typically stored in standard formats (e.g., Dbase) for easy maintenance by the users. For many years this caused a lot of tension between users of rival GIS systems who were unable to read each other's spatial data without going through a translation process to some common data structure. In the days of small disk storage capacity and small amounts of computer memory, these system-specific data structures often involved dozens of tables in a complex directory structure. The link between the table(s) containing the locational information about the features and the table containing the attribute information was critical, and if it got broken or corrupted, the GIS would not work well. You might know where something was but nothing about it, or you might have a lot of attribute information but no way to visualize it in relation to other features. One GIS had (and still does if you select the layer model) the interesting quirk where if you eliminate the feature from the map-like data view, the associated record in the attribute table is also removed. But if you remove the record from the attribute table, the locational information in the geographic table is not removed. So you end up with a feature about which you know nothing but which still appears on the screen.

127

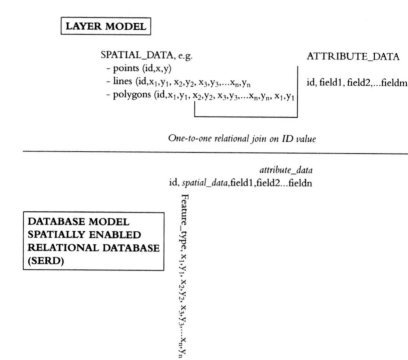

Figure 5.1   Layers and spatially enabled relational database.

In the database model all the data for each geographic feature is stored in a row in a table, but the spatial data are stored in an array (list) in one cell in the table. The data structures for these arrays differ among the DBMSs that support spatial data, and software designers continue to work on more efficient and useful ways to store spatial data in this data model. But whichever approach you select, you will still need to deal with issues of design for the fields in the table(s).

One reason that the design issues for attribute data are distinct from the design issues for spatial data is because there are usually many more people interested in the attribute data. Understanding the design issues for the spatial data is very important to the people who will be mapping and analyzing the geographic information, but understanding the attribute data is important to anyone who has to access the data to maintain or use it in any way, whether or not they ever view or use the geography. Designers of geographic databases need to recognize that there may be many users who will never view or make a map and will have no need to relate directly to the geographic information but will have to make use of the attribute information on a regular basis. Much of the input, analysis, and output activities in a geographic database have little geography involved.

You deal with design issues for attribute data on a table-by-table basis. Although the schema can be quite involved and the relationships complex, the structure of all the tables in the database is much simpler. Databases look complex at the highest level, but when you break them down into the component tables and relationships, the model simplifies considerably. A database table is a set of rows and columns where the rows represent objects or features and the columns, the attributes or things that we know about that particular object or feature. In a GIS these tables may be either geographic or nongeographic. Geographic tables have a field or fields that contain the locational information or a unique identification number field that allows the table to be linked to a geographic table; we will refer to them as G tables. Often, however, the database will often have tables that have no geography and contain information about nongeographic objects; we will call these D tables. For example, there may be a layer for the zoning polygons in a community where each feature in the table contains the geographic information (polygon outline, label point, etc.), a unique identifier for the feature, and the zoning category for that particular zoning polygon. Keep in mind that there may be many polygons representing a particular zoning category that are separated by areas with different zoning, so there will be more rows in this table than there are zoning categories in the community; there will be as many rows as there are unique, closed polygons of zoning areas. This G table may be related to a nongeographic data table (a D table) that contains only as many rows as there are zoning categories (see Figure 5.2). The fields or attributes in this table could contain a description of the category, permitted uses, a page reference to the zoning regulations where you find more information about the zoning category, and so on. Design issues for the fields in both types of tables are the same and by following some basic principles you should have a well-designed set of tables.

General Principles: Fields in Both D and G Tables

The following are the general principles that govern tables:

- Every table must have a *unique identifier* for each record, and it should be early in the table, preferably the first field. This location may not be possible for G tables because GIS software often reserves the early fields for spatial information. Most GISs actually create and manage a unique identification field for each feature in the table. In database terminology this field is a primary key, and each one in the table must be unique; no duplicates are allowed.

- Fields should *relate directly to the objects they model*. In the layer model this means that only geographic information would be in the G table and attribute information in the D table. A common mistake is to include too much information in the G table that properly belongs in a D table. In the example above, only a unique identifier and the zoning category are stored in the G table, with all the additional information in the D table. This eliminates a lot of redundancy in the database. If you stored all the information for, say, the R1 zoning category for each R1 polygon, the same data would have to be stored many times. The permitted uses are properties of the particular zoning category, not each separate polygon of that zoning category. In the Zoning_Polygons G table, the only field other than the unique identifier field, Poly_ID, is Zone_Cat_ID, which is the primary key in the D table Zoning_Categories. In the G table the Zone_Cat_ID field is known as a foreign key (i.e., a field in one table that is a primary key in another table and used for joining the tables together). Ideally, the fields in a G table should consist only of a primary key, foreign keys for table relationships, and attribute fields that relate directly to the spatial characteristics of the feature (e.g., length, area, perimeter, and x and y location).

♦ Each field in a table must have a *unique name,* but the same name may be used in different tables. For example, a field called Full_Address could exist in both a land parcel table and a table of owners of land parcels.

G Table: Zoning_Polygons

Poly_Id	Zone_Cat_Id
1	1
2	1
3	3
4	4
...	...

One-to-one join on Zone_Cat_Id

D Table: Zoning_Categories

Zone_Cat_Id	Zone_Category	Short_Desc	SFD_Perm	MFD_Perm
1	R1	Low-density residential	Y	N
2	R2	Medium-density residential	Y	Y
3	R3	High-density residential	Y	Y
4	LI	Light industrial	N	Y
...	...	...	...	...

Figure 5.2 G and D tables.

- Keep field names as *short as* possible and do not use spaces in field (or table) names; if you want to separate elements of a field name, use the underscore (e.g., Zone_Cat_Id) for unique ID values for zoning categories. Some RBDMSs have restrictions on the character length of a field. Use of both upper- and lowercase letters in the field make it easier to read.

- Make field names *descriptive* so that users will have an idea of what the field represents. Avoid too much use of numbers in field names. Most RDBMSs have a place for a field description where you can go into more detail than the name provides. Documentation of the field, its allowed values, length, and so on, is part of the overall database schema and also part of the metadata.

- *Avoid multiple fields of the same general type.* Take a pavement management example. If you want to maintain the names of the contractors responsible for work on a particular segment, there may be many different contractors that have worked on the segment for different purposes. A bad design is to include a set of fields such as Contractor_1, Contractor_2, and so on, in the segment D table because the relationship between contractors and street segments is many to many. Each segment may have many contractors that work on it, and each contractor may work on many segments. That would require a composite or link table in your schema to link contractors to street segments. But in a database representing people you might want to have fields that look like multiple entries but actually represent different things. A database might contain telephone numbers for each person to represent home, work, fax, and mobile telephones.

Specific Principles for G Tables

The following principles apply to G tables:

- There is always a set of fields in a G table that will be specific to the GIS you are using. When you create the table for the first time, the GIS puts these fields in. In some GISs these fields are invisible to the users, and in others they are visible but not editable. These often contain some spatial information such as area or perimeter for polygon layers and length for line layers. If the system uses the geo-relational or layered data model, there will be at least one unique identification field to link the spatial with the attribute information. The important principle to remember about these fields is not to alter them after they are created, losing the relationship between G and D tables. Table 5.1 lists the GIS-created fields for a table of polygon features in an ArcInfo coverage. The GIS will *maintain at least one unique ID number* that good to use as a primary key for this table.

Table 5.1 Fields in a G Table

Column	Item Name	Type	Explanation
N/A	FID	Object ID	Unique feature identification value. It will be numeric and sequential by 1 starting with the number 2. The polygon with FID = 1 is the background polygon necessary for creating topology. In some versions of this software the background polygon appears in the G table but not in others. This field cannot be edited.
N/A	Shape	Geometry	Geometric shape type of the feature such as polygon, line, point, region (multiple polygons), route (multiple lines), anno (text annotation). In this example it would be polygon for all the records.
1	Area	Floating point	Area of the polygon feature in the specified measurement units of the layer.
5	Perimeter	Floating point	Perimeter of the polygon.
9	Temp#	Binary	A duplicate of FID; also cannot be edited.
13	Temp-ID	Binary	Will usually be 1 less than FID and Temp#. This field can be edited if you wish to define your own primary key.

- *Foreign keys from other tables* that have a one-to-one relationship with the G table also belong in the G table. An example would be a PIN that linked the spatial feature to a record in a table containing assessment information. The PIN is the primary key in the assessment table and a foreign key in the parcel G table.

In the layer model include only fields required by the GIS, foreign keys from other tables, and other limited information that relates directly to the geographic nature of the feature. In the database model using a SERD, this is not a concern because both spatial and attribute data are included in the same table.

Generally the GIS creates and populates these fields for you except if you choose to create your own primary key. There is usually no reason to alter the values in these fields, and many users (and some GIS software systems) hide them from the user to avoid confusion.

Principles for Fields in D Tables

The fields you need to include in the data tables depend entirely on the applications you wish to support with the information. Most designers work backward, designing a report or form they wish to use and then adding the necessary fields to the appropriate data table. In some situations there may be some kind of standard template that sets up the fields for you or even fields the application is required by law to define in a certain way. But for each field you add to the table, you will have to define its type. Some RDBMSs and GISs allow for different types of data than others do and may call the same data type a different name. For example, some systems call textual data by the type name TEXT and others STRING or even CHAR, but they all stand for text data such as what you create on the computer keyboard. Finally, there are issues of compatibility between systems for data of the same type. For example, most RDBMSs allow a field generally known as a Memo field, which allows lengthy descriptions of objects with carriage returns and special symbols, but often the memo fields of one system will not transfer as memo fields in another. So the fields you need to define depend on your application, and the properties you can assign to those fields depend on your particular RDBMS and/or GIS. The need to have certain types of fields in your data tables sometimes constrains your selection of GIS. Fortunately, there are quite a few data types that are common to all systems as shown in Table 5.2.

There are design issues specific to each of these data types:

◆ *Numeric.* RDBMSs support many types of numbers, and you must specify exactly which type. Integer numbers allow a very wide range of values from large negative numbers through zero and to large positive numbers. Decimal points are not allowed in integer data. Byte numeric data is a special category of numbers restricted to the values between 0 and 255. They are extremely useful for identifying categorical data such as land use codes, building type codes, and so on. They are not good data type for unique feature IDs if there is a chance you would have more than 256 features in a table, as is very likely. With real, floating point, or decimal data (you will see all three terms used in databases) the concern is how much precision you need to the right of the decimal place (i.e., how many digits to allow for). With decimal numbers you must also remember when assigning the width for the field to account for the decimal point, which takes up a character width. Dates are actually stored and manipulated as numbers, but they may appear in reports and on input forms in quite different formats. A common date notation is YYYYMMDD, so May 2, 1999, would be 19990502 in the database. Date formats are quite varied, and the transfer of date information from one RDBMS to another is sometimes difficult. Additionally there are many different forms of currency notations available for numeric data that represents a monetary value.

Table 5.2 Attribute Data Types

Types	Subtypes	Description
Numeric	Byte	Restricted integer values 0–255; most efficient storage and retrieval.
	Integer	Variation in allowable range.
	Real (decimal)	May be single (decimal precision 7) or double precision (decimal precision 15). Some systems have subtypes with further precision, not usually needed in GIS.
	Currency	For use with monetary information. Many formats for different currencies.
Text (256 ASCII characters)		Also called string or character data.
	Fixed width	You must specify the width of the field; data longer than this are cut off at the specified width (truncation); 256 often the maximum possible width.
	Varying width	Avoids truncation and allows for longer text.
Date/time		Many different formats for international variations in how dates and times are used.
Yes/No or True/False		Sometimes called Boolean fields.
Memo		Text information allowing for multiple lines.
Automatic numbers		Numbers generated for unique identification purposes.
Object link embedded (OLE)		Link to other objects such as word processing documents, images, spreadsheets, etc.
Hyperlinks		Links to URL for Web access.
Binary large object (BLOB)		Image or sound files. Spatial data stored in SERD fields are often BLOBs.

- *Automatic numbers.* Usually the RDBMS or the GIS gives you no option in creating these automatic numbers; it does it for you in the data type that it wants to use, usually long integers.

- *OLE.* This type of object is a link to a document, spreadsheet, graphic file, or any object whose three-digit extension is recognized by your computer. Clicking on the information in the field, usually the path and name of the object, will open that file with the appropriate software.

This allows you to include spreadsheets, word processing documents, and so on, in the database. The object itself is not there, but the pointer to where it is and the link to the required software allow you to view the object from within the database.

- *Hyperlinks.* Similar to OLE objects, the object type has the URL for the object in the field, and clicking on it will open your specified Internet browser and open that object. Most GISs now have the capability to include this kind of object, so clicking on a point or polygon will take you to a location on the Web. Of course, there is still the problem of whether that particular resource is still where it used to be on the Web.

- In addition to determining the correct data type for each field in each table, you need to consider the acceptable values within the data type. With numeric and date/time fields you can usually specify a *domain*, which is a range of acceptable values for a field. For example, if a database was designed to process deed registration for a municipality, you could restrict the date for a field called date_filed to disallow any dates ahead of the current date. Or in a table that has a currency field amt_tendered you could disallow any negative numbers. If you had a field that used an integer to refer to eight possible conditions that an item could have, you could define the field as byte data and specify a domain from 1 to 8. This would prevent users from entering a value other one of than those integers. *Validation rules* apply usually to text data and can restrict values allowed in the field to a predetermined set. For example, if you were assigning land use categories to polygons and had a set of 15 possible text values that a polygon could have, you could set a validation rule to keep all entries to only those 15 text combinations. Domains and validation rules are critical steps in ensuring high quality in the attribute data in your GIS.

- *Blobs.* These are very specific to the particular RDBMS that is implementing them. Originally used to store image and sound files, they are now used in SERDs to store the coordinate information in data arrays.

Each field in the table must have its type and, depending on the type of field, other information such as width or precision explicitly determined. You will also be presented with a question about whether or not you wish to have an index created for a field. The primary key is an automatic index because it increases by 1 as each record is added. Because the data go into tables in no particular order, you have to search through the entire database, one record at a time, to find a record you want to retrieve. Without any indexes the system must sift through the records one by one until it finds the desired record. An index is simply a table that is ordered from low to high values on a particular field in the original table. The only other field in an index table is the primary key field from the original table. So if you search frequently by last name and create an index on that field, the records in the index table will be organized alphabetically. With an indexed search the index table is searched instead of the original table. By comparing the requested value with the value retrieved from a particular record in the index table, the search process can determine if the requested record is above or below it in the table. Figure 5.3 shows the process, which is easier to visualize than to describe verbally.

By jumping the correct direction in the index table halfway between where you started and where you ended up, you can reach the requested record in seven record reads or less. That record in the index table is joined to the original table on the original table's primary key field, which is a foreign key in the index table, so the requested record can be retrieved quickly from the original table through use of the index table. This rapid search process is independent of the size of the table (i.e., you find the requested record in seven jumps or less whether there are 2,000 or 2,000,000 records in the table). That is how an indexed attribute search works, and the next question is what field or fields should be indexed. Composite indexes can be created from multiple fields also. There might be a tendency to index on all fields, but that can actually slow down searches; therefore, create indexes only for those attribute fields you search on frequently. Indexes are important for spatial data as well and spatial indexes are discussed in chapter 4 – Spatial Indexing.

The sections above dealt with design issues for the schema, spatial data, and attribute data. In addition to design issues around the G and D tables, basically the left-hand side of Figure 5.4, which these sections have concentrated on, there are other elements of the schema that you will need design. These elements relate to input, management, application, and output activities.

Indexed search for P

ABCDEFGHIJKLMNOPQRSTUVWXYZ

Random jump to J. P is above J, so go halfway between J and Z, which is R. P is below R, so jump halfway between R and J, which is N. P is above N, so jump halfway between N and R, which is P. Search complete with four record reads. A sequential search from the top would have required sixteen record reads.

Figure 5.3 Indexed attribute search.

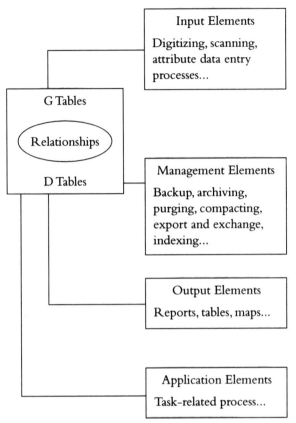

Figure 5.4 Schema design elements.

Designing Input Elements

Input elements are applications or processes you will use to get spatial and attribute data into your GIS in the appropriate fields of the appropriate tables. For attribute data these elements are typically called input *forms*. In a GIS that is replacing a previous paper operation that used paper forms for data input, many people attempt to redesign the paper form to fit on the computer screen, but this often does not work because of the relatively limited space on the screen. If you expect lots of attribute data modification — editing, adding, and deleting — in your GIS, the design of the input forms will either dramatically speed up or slow down the process. Pull-down menu options can minimize typing errors, and good validation rules are a necessary part of the table definition. Most nonspatial users of the GIS, and there will be many, interact with the database through input forms and output reports. It is important to consult them about the arrangement of elements on input forms and about how each input is related to the table structure of the schema. Forms also change, and users discover needs for new forms all the time, so the maintenance and update processes for input forms are ongoing. Because the table structure is relational, forms will often access and/or update data in more than one table, so there are control and access issues in the design of forms. No user should have access to a form that will allow that user to modify data the user is not supposed to modify. This is exactly the same in the paper form world where only certain users have supplies of certain forms and the authority to fill them out. And, like paper forms, data input and modification forms can be reviewed and approved by supervisors before being committed to the database. Most of the concern over electronic signatures involves electronic commerce and legal paperwork, but input forms in complex relational databases often require multiple signatures, and as the legal issues get resolved will be part of the design process for input forms.

The design of input forms is important not only because poor design can lead to inefficient data entry; it can lead to data corruption as well. When addresses are entered on a form for a community that might have several thousand different street names, you will want to discourage (or disable) the actual typing of the street name because of number of errors that will occur. Therefore, you might want to design a system where the user clicks on the letter of the alphabet the street begins with to produces a pull-down menu for selecting the street. Boxes on the screen can be constrained, through validation rules for the table column, to accept data only in a particular format or range. The input form can automatically place dashes between the elements of telephone numbers and even produce lists of common area codes but allow for input of area codes not on the list. Input forms can also be set up to turn whatever text is typed in into uppercase text so the user doesn't have to remember to hit the Caps Lock key on the keyboard. When a form on a computer screen is filled out from a paper form, it is important to design the computer screen form so that the user does not have to jump from place to place to find the corresponding information on the paper form. Although there is less and less direct attribute entry from paper forms (they have been replaced with direct database access), there are still many applications where a paper form is filled out in the field and entered into the database later. This paper form then also becomes an important backup record of the transaction or information. Finally, the design of many forms that might be used for input into a GIS may have legal requirements for what must be on the form, so that exact format must be used. Input forms are significant design elements for a GIS, but because they very so widely from application area to application area, we must leave them for now.

Design of Output Elements

Beginning users of GIS think that most output from a GIS is a paper or computer screen map of some location or set of locations. However, there are many different types of possible output from a GIS, and each has its own design concerns.

Map Design Issues

Map design is a complex issue and really beyond the scope of this book. But it is clearly important, or there would not be some sort of map design competition at almost every GIS conference or convention ever held. The ability to display information visually is perhaps the most important way in which a GIS differs from a nonspatial information system.

Cartographers and map designers disagree even about the absolutely required elements of a map. For example, some will insist that all maps must have an arrow of some sort indicating true north, whereas others might say that is required only if north is not parallel to the vertical edges of the paper. Some feel strongly that all maps should have at least one neat line that surrounds all the design elements on the map. All do agree that maps needs titles, legends, and some indication of scale. Each of these map elements, in addition to the base geographic information, has its own design concerns. For each type of textual information you need to decide on type font, style, size, spacing, alignment, and so on. To show scale you need to decide on representative fraction (1:35,000), linear scale bar, or scale statement (1 inch equals 1.5 miles). Even the decision of whether or not to use a comma to separate thousands is a design question. And then there are the questions about how to arrange all the elements on the map. Does the title go across the top or the bottom or perhaps the side? Where does the legend go?

Issues of map design become even more important when routine map output is a planned part of your GIS. For users who need these maps to make decisions or go into the field, the map really is a surrogate for the entire GIS; without the mapped output, the GIS is of limited use to them. Fortunately, GISs come with functionality that allows users to create standard templates into which different data can be placed. Some even have system-specific programming languages that allow you to define exactly how you want a particular map to look and where the map elements will be placed. To make the map may require a little screen input, and the program will generate the map.

One choice you must make is where you are going to store the information on map design. At the very simplest level you can print out how you want a standard map to look and mark the font types, sizes, colors, line widths, and legend contents with a pen on the map. This seems rather crude, but it works; it allows multiple users to produce maps that appear very similar and provides, very easily, a consistent look for your mapped output. One step up from that is to store the information in what are often called *map templates.* Templates are blank maps that are already sized and symbolized according to your design criteria, and new information is plugged into the template. Finally, some GISs have scripting or programming languages that allow extremely close control over mapped features. Using these programming languages you can locate text exactly where you want it to be, change font types, sizes, and so on. Just by changing the names of some layers, some text for titles, and other small changes in the program, the maps will come out looking exactly the same every time without having to set any of the parameters about the mapped information. This kind of programming is very straightforward and does not require high levels of programming skill

One of easiest ways to develop ideas about map design is to look at other people's work. Most GIS software systems publish regular newsletters or even magazines that highlight how organizations are using the software, and they usually have good examples of well-designed maps. GIS conventions and conferences are also good areas to look for interesting designs. It is difficult to overestimate the importance of high-quality mapped output from your GIS, especially with top management. Tangible products that people can hold in their hands, such as maps, help make your complex database more real to people. Relational joins, attribute definitions, and database schema are excellent tools to demonstrate to other database professionals that you have a well-designed system, but a high-quality, large color map or set of maps does the same thing for less sophisticated users.

Map design is a creative process that has an information foundation. The information the map presents must be correct, and the design of the map must present the information in such a way that the users find it useful and possibly even attractive. Not all GIS users or designers are skilled at map design, and if your GIS installation leads to large amounts of mapped output that may be seen by professionals outside your organization, time spent on map design is never wasted.

Application Design

The applications are where this process all started (see Figure 1.1) because they are the critical element of a GIS, the element that drives the rest of them. So it is only natural that there would be some design concerns around applications. Applications or processes that users execute to interact with the GIS often contain other design elements, and you can think of the application as the container for these elements. It may have several input forms, a management section where the data go to supervisors for verification and final insertion into the database, and a generated report that is filed separately as a paper backup of the transaction or update. The most common way to design and visualize applications is through construction of flow charts, or diagrams, that show how the process moves from the user's perspective. These flow charts work as a blueprint for the programmers or software designers who have to make the application operational. Although the focus here is on applications that are programmed to take the user through the process, they can work just as well as descriptive set of procedures the user has to implement to obtain the desired results. In this case the application is a set of instructions. So either way, they are instructions for the user or the program with input from the user. They can be complex or relatively simple, but at a minimum an application design for a GIS application should have the following components:

- A list of all required tables and the fields from those tables for the application. If there are many-to-many relationships that require composite tables, make sure those tables are available as well.

- Diagrams of input screens and output reports, including map-based forms and reports. If the application is automated with scripting or programming, these forms and output reports will have to be quite explicit.
- A flow chart of the process with decision points clearly marked. These are places where the user needs to make a decision or supply some information to the application.

Figure 5.5 is a simple version of a zone change notification process that a municipal government might have to perform. The final decision point was added after discussion with a staff person who did notifications such as this because they often include a few extra parcels that just look like they ought to be notified, even if they do not explicitly match the stated distance criterion. This application would also involve programming to allow the removal of the parcel for which the change is being requested from the set of parcels that are within a distance of 500 ft from that parcel. This is because that parcel itself will always pass that query; that owner is notified differently of the hearing than the nearby neighbors are. The principal output of this application, which would also have to be designed, is a set of form letters that go to the property owners. In most GISs it is possible to directly create these form letters, but you may wish to export the information to a spreadsheet or database table for merging in a word processor. This application could be easily extended to provide information to a management report about when the letters were generated, who did the work, and how the responses were tracked.

How They Did It — Kansas Geospatial Data Addressing Standard

Addresses are so important and nonstandardized that the state of Kansas set up a committee of 15 people that had five meetings over a year and a half, presented its results to a GIS forum, and finally published the standard 20 months after the first meeting. As usual, most of the work was done by a handful of people, but the time spent and variety of participants gave it some importance. The standard rests on a single table, the master address table or address repository and the relationships between that table and other tables. This table, "a geospatial dataset of situs address points" provides a single point for every address. The standard is very specific about the components of the address, with all the components of the address being alpha or text information except possibly the unique identifier for the address. Such specifics as "Numeric street names shall be written using numbers rather than spelled out. For example, '1ST' is to be used rather than 'FIRST.'" This is the level of detail needed in an address standard. It also incorporated direct reference and use of other standards such as the *U.S. Postal Addressing Standards, Publication 28* and the Planning Advisory Service *Street Naming and Property Numbering Systems, Report No. 332.* This is along the lines of not reinventing the wheel.

continued

Although the specifications for the structure of the master address table are very detailed, the specifications of the relationships between tables are less so because there is so much variety in the other tables that might be linked. Their major concern was the linking of land parcels to addresses because a single parcel may have many addresses. So the next question became: How do you link a geographic table that uses parcel identification numbers with the nongeographic master address table? They wanted to avoid multiple records in the address table, which should have only one record per address. They created a parent-child relationship in the master address table by adding a second column to store the ID number of the primary address. They presented some suggestions for a structure of an address database, shown in Figure 5.6, and how you might link it to a crime incident table, a CAMA table, and others.

The purpose of the standards was principally to assign addresses to land parcels; in Johnson County, Kansas, for instance, on the southwestern edge of the Kansas City metropolitan regions, there are 19 different entities that assign addresses to land parcels. The 911 emergency dispatchers and managers were particularly interested in how these standards work because their goal is to eliminate as many duplicate addresses as possible for public safety reasons. The standard was not intended to be the last word in dealing with addresses in a GIS, and the drafters expected extensions to the design to include dummy addresses for buildings, vacant lots, parks, and so on, for emergency dispatching purposes.

There has been some adoption and use of the standards statewide, according to Jay Heerman, Johnson County Address Coordinator, and the process isn't over. They hope to demonstrate the standards to other parts of the state that are having the same difficulties and to help create standard addresses that can be easily searched and located.

There are many excellent texts and resources for design of RDMSs, and database management systems come with many tools and resources for schema design. But all design considerations go back to your careful assessment of the tasks or applications your database has to support. Without that step you are likely to leave out critical tables and fields that you will need to support an application. This will erode confidence and support for your GIS. You should design a database that meets you needs but avoid the temptation to overdesign it. The design should be flexible enough to incorporate new tables and processes but tight enough that you are confident you have high-quality data in your tables. And like most design processes, good GIS database design is a blend of technical skill, creative crafting, and thought. Time spent on design is never wasted.

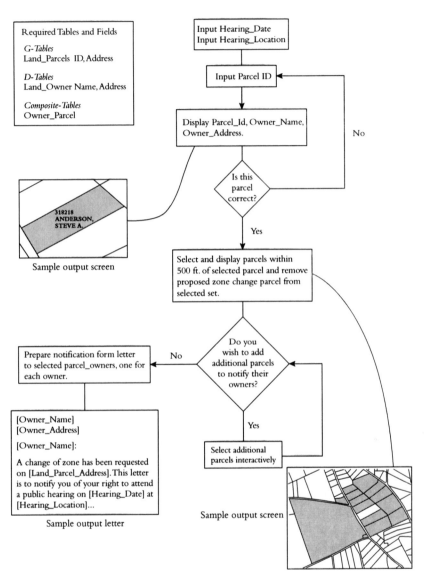

Required Tables and Fields

G-Tables
Land_Parcels ID, Address

D-Tables
Land_Owner Name, Address

Composite-Tables
Owner_Parcel

Input Hearing_Date
Input Hearing_Location

Input Parcel ID

Display Parcel_Id, Owner_Name,
Owner_Address.

No

318218
ANDERSON,
STEVE A.

Sample output screen

Is this parcel correct?

Yes

Select and display parcels within
500 ft. of selected parcel and remove
proposed zone change parcel from
selected set.

Prepare notification form letter
to selected parcel_owners, one for
each owner.

No

Do you wish to add additional parcels to notify their owners?

Yes

[Owner_Name]
[Owner_Address]

[Owner_Name]:

A change of zone has been requested
on [Land_Parcel_Address]. This letter
is to notify you of your right to attend
a public hearing on [Hearing_Date] at
[Hearing_Location]...

Sample output letter

Select additional
parcels interactively

Sample output screen

Figure 5.5   Application design elements.

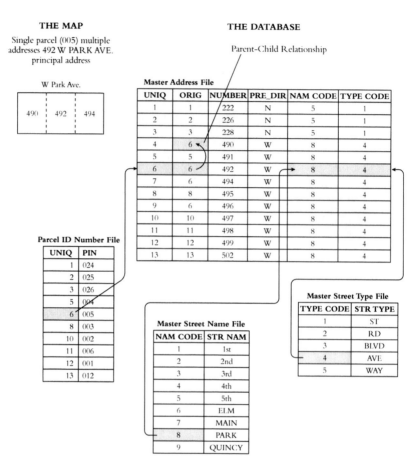

Figure 5.6 Kansas addressing standard, data links. Adapted from Kansas Addressing Standard Committee, 1999.

ADDITIONAL READING

Begg, C., and T. Connoly. 2001. Database Systems: A Practical Approach to Design, Implementation, and Management, 3rd ed. Addison-Wesley: Boston.

Date, C. J. 1999. An Introduction to Database Systems, 7th ed. Addison-Wesley: Boston.

Halpin, T. 2001. Information Modeling and Relational Databases: From Conceptual Analysis to Logical Design. Morgan-Kaufman: New York.

Harrington, J. 2002. Relational Database Design Clearly Explained, 2nd ed. Morgan Kaufman: New York.

Hernandez, M. J. 1997. Database Design for Mere Mortals: A Hands-On Guide to Relational Database Design. Addison-Wesley: Boston.

Rigaux, P., M. Scholl, and A. Voisard. 2001. Relational Databases: With Application to GIS. Morgan Kaufman Publishers: San Francisco.

Riordan, R. M. 1999. Designing Relational Database Systems. Microsoft Press: Redmond, WA.

Silberschatz, A., H. F. Korth, and S. Sudarshan. 2001. Database System Concepts, 4th ed. McGraw-Hill: New York.

INTERNET RESOURCES

Brief discussion of the integration of a geological database with a spatial map database (Kentucky):
pubs.usgs.gov/openfile/of99-386/anderson1.html

Remotely Sensed Data as Background Layers and Data Sources

Geographic databases are supposed to be models of the real world, but they often end up as models of maps. To be sure, a GIS is a single map that covers your entire area of interest, and you can turn layers on and off, add and remove layers at will, rearrange layers, resymbolize layers, zoom, and pan, and this has many advantages over paper maps. But points lines and polygons against the white background of a computer screen look, feel, and behave more like a map than the real world. So instead of a representation of reality, you end up with much better, intelligent maps. This is certainly a step forward, but it is possible to bring more real-world context into your GIS if you have a need for it.

A common method for doing this is to have some digital and visual representation of what is on the earth. This is usually an aerial photograph or some processed digital satellite data. When you have a need for this background visual information about the earth's surface, you must work at the intersection of three distinct but increasingly related disciplines: photogrammetry, remote sensing, and GIS. Photogrammetry is the art and science of representing and measuring the real world through photographs, usually taken from airplanes. These photographs may be analog (film) or digital and may display the world in black and white or in color, sometimes in infrared radiation. Remote sensing is measuring reflectance of the earth using sensors sensitive to specific regions of the electromagnetic spectrum. This digital data can be processed in many ways to highlight certain earth features. These disciplines evolved independently and in advance of GIS, but their products can be easily used in a GIS if you take the appropriate steps in the planning and design phases.

How would you know whether you will benefit from the inclusion of digital background data, either aerial photography or remotely sensed data? Some applications gain little benefit from such information. For example, if the principal use of your GIS is vehicle routing from address to address, having detailed images of buildings, trees, fences, and so on, is no real value and only slows the system down. So a GIS application used to route emergency responders to a location might not need this kind of background. Once the responders reach an address, there might be a real value in having an aerial photograph behind the data. In one town a police officer was responding to a breaking and entering call and was dispatched to the next street to approach the house from the rear. Because there was some urgency, he was running and tripped over a low fence that he could not see in the dark. As they reviewed the incident with the GIS using a 0.5-ft-scale aerial photograph image as a backdrop, the fence was clearly visible. A dispatcher with access to that information could have warned the officer, and he could have avoided the injury. Further, it is our experience that even in applications that do not require the presence of a backdrop image of the real world, users prefer to see that sort of backdrop and have higher levels of confidence in the data when they see the correspondence between the points, line, and polygons in the GIS and what they represent on the surface of the earth.

Aerial Photography as Backdrop Information

Aerial photography converted to a digital orthophoto image is a common data source for in many municipal and utility GIS applications. Photographs from the air are taken from a range of platforms including telescoping masts, radio-controlled blimps, hot-air balloons, kites, helicopters, and airplanes. Most of the low-elevation platforms are used for oblique photography to show individual buildings and small areas rather than as backdrops behind GIS data covering larger areas, so we do not consider oblique photography in this section. Municipal and state governments take aerial photographs at intervals of 5 to 10 years, if they can afford it, and have used the printed-paper output of the photographic process for many planning and management applications. Water, gas, and electric utilities are also big users of aerial photography as is any organization with widespread land management responsibilities such as forestry and range management. For the past 30 to 40 years aerial photography has been the source data for most visible mapped data — roads, building outlines, areas of vegetation, sidewalks, driveways, utility poles, fence lines, and so on, and, by using stereo pairs of photographs, for contour lines. Before organizations began using GIS, the photographs were the source of their paper maps. They are still the standard source for the federal government's mapping activities. Capturing information through photography is, per square mile, much less costly than on-the-ground surveying. Today, aerial photography, corrected for terrain and camera tilt to form an aerial map, is the source for the same information, except instead of drafting it onto paper maps or producing photographic prints, the data are prepared digitally for inclusion in a GIS. These digital views of the real world have not completely replaced the paper prints of the film negatives, and many organizations use both products for different purposes.

Beginning users of GIS assume that they can simply scan existing paper aerial photographs using a desktop scanner and then quickly drape their vector information on top of the scanned digital image. If it were that easy, everyone would do it, but it is not, so a good idea is to obtain some expert advice before planning use of aerial photography in your GIS. Most civil engineers have some training in this area, and even in this increasingly satellite-oriented world there are many photogrammetrists and aerial photography firms around, most of whom now have experience in producing output for use in GIS.

The best time to plan for the integration of aerial photography into your GIS is when you are beginning to design the photography process or preparing the RFP. It is becoming increasingly common for aerial photography firms to provide digital versions of the photography along with the traditional paper prints and film negatives, but if your organization wants that, you need to incorporate the digital version specifications at the beginning. Preparing aerial photography to be visual backdrop to vector GIS data are much more difficult when you work from the final paper photographs; it is orders of magnitude easier to do it simultaneously with the production of the paper products.

Our discussion of aerial photography is not going to involve the myriad technical concerns but will rather concentrate on the questions of importance to GIS. We include some of the simpler equations so that you can understand the basic concepts involved, but there are many more technical questions that users must consider before using aerial photographs as backdrops to their vector data. (the equations presented are from Lillesand and Kiefer 2000). We also expect that you will want your photography processed into digital orthophotographs. An orthophotograph is processed from original scanned data that have been geo-referenced to a coordinate system and draped over a digital elevation model of the area that contains the elevation information. The goal is to produce a map-like image, which has been corrected for distortions caused by the lens, camera position during flight, and variations in terrain. Because locations at higher elevations are closer to the camera than low-lying spots, the scale is not constant. In hilly terrain a straight-line feature such as a road or power line will appear to waver in a standard aerial photograph. After processing as an orthophotograph, it will appear as the straight-line feature it is. As an analogue or paper product, orthophotographs are produced through use of a direct optical projection orthophotoscope, but for inclusion in a GIS it is more commonly done with digital computer processing. The mathematical processing is quite complex, and it does require a digital elevation model representing the terrain of the area in question. Usually, this information is gathered at the same time as the photography, but it is possible to use existing DEMs as the terrain-correction data source. The mathematical calculations involve the averaging of pixel values, so some direct information is lost, but the result is a digital photograph in which it appears that the camera was pointing straight down over the entire photograph. This is basically aerial photography processed to look like a map. You can then draw your GIS vector information on top of this map-like image (see Figure 6.1).

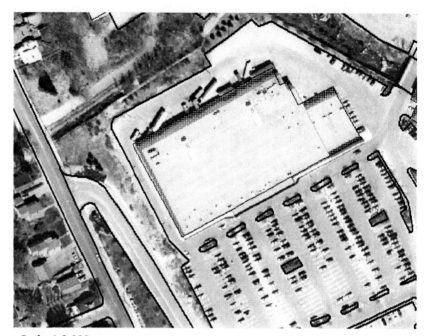

Scale: 1:2,000
Cell Resolution: 1ft
Gray Scale Resolution: 256 levels
Source: SBC - Southern New England Telephone
(used with permission)

Figure 6.1   Digital orthophoto with vector overlay. Scale: 1:2,000; gray-scale resolution: 256 levels. *Source*: SBC-Southern New England Telephone (used with permission).

Color or Black and White Photography

All the decisions that an organization needs to make around using aerial photography in a GIS are important, and the choice of color versus black and white is one of the first to make. Color and black and white infrared photography, which use film that is sensitive to the infrared range of the spectrum, is also possible, but these film types are used principally in vegetation studies. Digital color infrared imagery can be processed to produce something close to true color, however.

In choosing between color and black and white you need to consider issues of expense and the visibility of features on the ground. Color aerial photography, for the same size area at the same scale and resolution, is several times more expensive to fly and process than black and white photography. Additional prints from the color negatives are equally more expensive. The other key concern is whether you will be able to visualize the needed features from color photography. In areas that are principally covered with trees, and this can include many urban and suburban regions, the preponderance of leaf cover will hide almost all the features you might want to capture. This is why most aerial photography must be flown when leaves are off trees. Generally, black and white photography will provide higher contrast and allow for clearer distinctions between natural and artificial features. Areas that are almost completely developed with little natural cover, however, are particularly appropriate for color photography. The recent aerial survey of New York City, NYCEMAP, was flown in color, and the photographs show considerable detail. Color is also good in regions that are not heavily treed. Color images appear more real, but often more information can be seen on black and white images. And once the orthophotograph is in digital form, many different types of enhancements, such as improving contrast and altering brightness, will make certain features stand out. Digital color photographic imagery is usually gathered through three sensors that record intensity of color in different bands (red, green, and blue) of the electromagnetic spectrum. These bands are combined to produce a single color composite image. By varying the importance of the three bands in the production of the final image, it is possible to make certain types of features stand out; allowing each band to be equally important produces the image closest to real-world color. You can also color orthophotographic imagery to approximate black and white coverage. Generally you get more options with color, but the additional acquisition, processing, and storage costs are significant, so many organizations continue to use black and white orthophotography as a GIS backdrop.

Another issue of resolution for both color and black and white digital imagery is the level of resolution you want for the color information. It has become almost standard to use a 256-value gray-scale range for black and white data where the value 0 indicates pure black and 255, pure white. The human eye really cannot distinguish that many levels of gray but can at the levels below it (128 and 64, the next lower powers of 2). Using the powers of 2 (256 is 8 bits, or 2^8) is a function of computer storage. Because 8 bits equals a byte, it is easy to calculate storage requirements for black and white data as well because each pixel will require 1 byte of storage space. Color resolution is usually has 256 levels as well, with three different bands, red, green, and blue. The number of possible color combinations with three bands of 256 values is huge and produces many more colors than human eyes can distinguish.

Scale of the Original Photography

Anyone who has flown in airplanes at different elevations is quite aware that the closer the ground you fly, the smaller the features you are able to identify. At 30,000 feet it is difficult to spot individual cars on a highway, but at 10,000 feet you can identify them by color. In aerial photography the scale of a photograph is determined by the focal length of the camera (distance from the shutter to the focal plane), the height above the ground of the camera, and the height of the terrain being photographed. For a given focal length camera, as the airplane flies higher the scale gets smaller, the equation is as follows:

$$S = f/H \_ h'$$

where S = reciprocal of scale (e.g., at 1:24,000, S = 2,000); f = camera focal length (measured in meters; a 200-mm focal length lens would be 0.200); H = height above the ground (meters); and h' = height of ground above sea level (meters). For example, to obtain photography at scale of approximately 1 inch to 200 ft using a camera with a standard focal length of 152 mm flying over a flat area at sea level would require flying the plane at a rather low level of 370 m (1,220 ft) Using a focal length of 0.210 mm would allow you to capture data at this scale from 500 m (1,650 ft). Scale on an aerial photograph is not constant over hilly terrain because some of the area is closer to the camera than other portions. Also, the larger the scale you require, the more photographs will have to be taken and processed.

Two primary standards are used to determine the accuracy of the mapping products derived from a photogrammetric project. The first is the National Map Accuracy Standards (NMAS) published by the USGS, and the second is the National Standard for Spatial Data Accuracy published by the American Society of Photogrammetry and Remote Sensing (ASPRS) (See chapter 5 — Accuracy, Precision and Completeness). Both of these standards are very clear on the measurements and procedures that can be used to test the resulting data products for a mapping project, but they both leave much room for interpretation by a photogrammetrist. For example, to be in compliance with the NMAS standard and produce a final mapping product of 1 inch =100 ft, the height of the aerial flight can vary from a photo scale of 1 inch =600 ft to 1 in =1200 ft. The test to measure the accuracy of the data is that 90 percent of the features collected are within one-fortieth of the mapping scale, or 2.5 ft. There is nothing in the standard that defines what the scale of the flight needs to be to produce this result; thus companies that provide these services are left to their own discretion as to what scale to use.

One of the disadvantages of digital orthophotography over traditional film aerial photography is the limited range of scale to which you can zoom. The silver halide crystals in the film negative that contain the level of darkness in a photograph are extremely small and vary in size. This means that you can develop prints from the negative at a scale much larger than the original photograph and still have considerable detail in the image. Eventually small features get fuzzy, but you can blow up or zoom in very far using photographic negatives. If you process the photography digitally for inclusion into your GIS, the photographs will be appropriate for viewing only at certain scales. This is determined by the cell resolution you select for the digital information. The smaller the cell sizes the smaller the features you can recognize. Typically digital orthophotographs use cell resolutions of 1 m, 1 ft, or 0.5 ft. If you zoom in too close to the photograph at a large scale, the pixel structure of the data obscures any features you might wish to identify (see Figure 6.2). Most GISs allow you to control the drawing of layers and restrict that to a specified scale range. So if your area of interest were large and required several dozen photograph tiles to cover it with imagery, you probably would not want all that imagery to draw when you were looking at the entire area. Rather, you would want the imagery to appear when you were zoomed in to a certain scale. But if you zoom in too far, the pixel structure of the data will be apparent, and it will not be possible to discriminate features on the image. Setting the scale range for orthophotograph drawing is a design question, and the smaller the cell size resolution the closer in you will be able to zoom.

Scale: 1:2,000
Automobiles recognizable,
sidewalks and frontwalks
visible.

Scale: 1:500
Not much change; same
features still recognizable
but overall image is grainier.

Scale: 1:100
Pixel data structure clearly
visible, truck barely recognizable.

Figure 6.2 Scale effects in digital orthophotos.

Ground Resolution

The scale of the photograph is related to the height of the flight and the focal length of the camera and also helps to determine the level of detail you can possibly distinguish in the resultant photograph. It is very important to know what the ground resolution is of the photograph because this determines the smallest feature you could possibly distinguish. In a GIS designed for an electric utility you will want to be able to distinguish the tops of utility poles from the surrounding area. Even if you mount a reflector on the top of the pole, you are asking an interpreter to identify a circle 12 inches or so in diameter; that is quite small. The amount of spatial resolution you can derive from a particular photographic system is impossible to quantify because there are so many factors that affect it. Atmospheric conditions, film developing conditions, the contrast level of the film, the speed with which the film accepts light, vibrations to the camera; the list is long. Most factors are difficult to quantify, but there are some that can be measured to help determine what level of detail you can capture. These are the scale of the photograph and the resolving power of the lens and film combination. Although there are sophisticated ways to measure resolving power, the final measurement consists of the number of alternating black and white line pairs that can be distinguished per millimeter of film. This resolving power for modern films and lenses is tremendous and is what allows the identification of relatively small features on the landscape. Assuming that there is no degradation of ground resolution by the non-quantifiable factors, and there will be, ground resolution is as follows:

$$\text{GRD} = \text{reciprocal of photograph scale/system resolution}$$

So if the resolution of the system of lens and film were 40 lines/mm, and the scale of the photograph was 1:15,000,

$$\text{GRD} = 15,000/40 = 300 \text{ mm, or } 0.30 \text{ m}$$

This is the optimal level of detail you could resolve on the photograph, but you will rarely attain it. What you can actually see is affected by the visual contrast in the landscape, distortions caused by camera vibration, air quality, film transport, and so on. The ground resolution provides a theoretical minimum of the size of feature that can be distinguished from something else.

If a cell resolution of 1 m is adequate for your needs and you are not too concerned about the regency of the imagery, the most inexpensive way (in the United States, at least) to incorporate image data in a GIS is to use the digital orthophotographs provided by the USGS (digital orthophoto quarter quadrangles or DOQQ). Each image covers one-quarter of a 7.5-minute quadrangle at a scale of 1:12,000 with a cell resolution of 1 m. For urban applications the scale is too small, but for larger areas of interest the images can be a useful addition to a GIS. They can be obtained at very low cost from the USGS, many are on the World Wide Web, and there are several resellers and packagers who will enhance and combine these images at a reasonable cost. Distribution is on CDs with one county per disk, and its very affordable pricing policy makes this data source very inexpensive. Coverage of the United States is not yet complete, and similar resources are not available at such low cost in the rest of the world. The largest problem with using this digital orthophoto source is that the images are usually several years old and the update cycle is long. There is a tendency to think that the image behind the data is representing the world as it is today, but it is always at least 6 months old and usually much older than that. It is a little odd to see an up-to-date map of land parcels that clearly indicate a subdivision draped over the farm that was there when the image was taken. CDs from the USGS come in the UTM projection, so you would need processing capability to match it with your selected projection.

Number of Required Photographs or Tiles

The number of photographs you will need to take is a function of the size of your area of interest, the scale and related ground resolution you have selected, and the amount of overlap you need between adjacent photographs. The standard frame aerial camera takes a negative 230 mm on a side, approximately 9 inches. So at a scale of 1:15,000 a single photograph would show an area of 11,250 square feet, or 2.1 miles on a side for an area of 5 square miles. If you are going to use the photography to derive contour lines and elevation information, you will need to have sufficient overlap in the photographs so that stereo imaging is possible. Even if you are not going to do that, and most do, you need overlap because of scale and direction distortion that occurs in photographs caused by the camera's lenses. This distortion reduces the usable area of a photograph. The airplane flies the region in approximately straight lines of flight, and common overlaps in the direction of flight are 55 to 65 percent, with 30 percent between adjacent flight lines. This increases significantly the number of photographs required to cover a given area. Determining how many photographs to take, at what elevation, with what focal-length camera and on what kind of film are all technical considerations that relate directly back to your determination of the smallest feature you wish to be able to recognize on the final photograph. The contractor will be able to calculate these factors and their costs, but only the final user can determine the level of detail that is necessary for a given application.

Capture Data as Well?

Three primary types of data are typically collected using aerial photography for inclusion in a GIS: planimetric features, topographic features, and digital orthophotographs. Planimetric features are those natural or artificial features that can be distinguished and are visible on the surface of the earth. Typical planimetric features include the following: edge of pavement, sidewalks, driveways, parking areas, structures, pools, tree outlines, brush/shrubs, street trees, signs, parking meters, flag poles, utility poles, lakes, ponds, streams, rivers, wetlands, and bridges. Decisions about which features are going to be captured from the photography are made before the photography is flown because it is a key determinant of the scale of the photography. The smaller the feature you want to obtain for your GIS (e.g., parking meters), the larger the scale of your photography.

Digital orthophotographs are the principal source for planimetric data used in small- and medium-scale GIS applications. Because of their map-like appearance, it is possible to use heads-ups digitizing and prepare vector layers from visible features. The selected cell resolution determines what features you can see on the photograph, but the decision of what features to capture in vector format off the photographs is a separate one that also has cost implications. Road centerlines, rights of way, building outlines, hydrographic features, fence lines, walls, elevation contours (if sufficient overlap was allowed for), pavement striping, street furniture, driveways, and sidewalks can all be seen on 1-ft or better aerial photography. The capturing of the vector data is done at the same time as the images are processed, and you may specify as much or as little vector information as you wish and can afford. The derived vector information and the orthophotographs are separate items and have different uses. One GIS manager in a Maryland County surveyed his users about their preferences, asking them if they wanted updates of just the photography or of just the planimetric information and was surprised that they choose to have the photographic updates. The incorporation of these images brings much of the real world into analysis.

Topographic features are those that represent the vertical elevation of a surface in a three-dimensional plane. These features are typically referenced by the contour increment that they represent. For example, common increments are 2-, 5-, or 10-ft contour increments where each line represents the corresponding change in vertical elevation on the surface of the ground. For a 2-ft contour interval, each line on the map would represent 2 ft of vertical elevation change. The other common topographic features found are spot elevations; they are specific elevations that are found at high pints, low points, and saddles on a topographic map.

With regard to standards applied to typical topographic maps, the vertical accuracy of a map is typical defined as one-half the contour interval. So for a 2-ft contour map, the accuracy of any point on the map would be ±1 ft. The contours are developed from overlapping stereo pairs of photographs and known elevation points, and you must select the contour interval before the flight.

Recently a new technology for gathering topographic information has been developed, light detection and ranging (LIDAR). Simply, an air-borne LIDAR system, on the same plane with the aerial cameras, sends a laser pulse to the ground and measures the time it takes for the reflected pulse to return. Because the speed of light is a constant and the elevation of the plane can be determined when the pulse was sent, the time it takes to receive the reflection will allow the system to calculate the elevation of the point of reflection. With pulses being sent hundreds of times per second, LIDAR returns a very complex picture of topography. For example, in heavily treed areas you will receive some reflection of the pulse from the top of the canopy, some from the understory layer, and some from the ground. LIDAR will return rudimentary building outlines. Because many pulses will reflect off a building, the flat roof will be visible and the spaces between buildings as well. LIDAR returns a very detailed three-dimensional picture of the landscape and what is on it and requires quite a bit of postprocessing analysis for interpretation. LIDAR has many uses in measuring distances, speeds, rotation, and chemical composition and is heavily used in atmospheric research and volume estimation. In GIS the principal use is to provide detailed topographic information either in a high-density, small-pixel raster layer or as processed contour lines. Some exciting applications are being developed for LIDAR data and GIS. One is using LIDAR to update existing data layers. A shaded-relief LIDAR image with building outlines drawn over it will show buildings that are no longer present. The outline of the building in the LIDAR image will not look exactly rectangular because of the way the pulses are sent, but the absence of a building underneath the outline is evident and the outline can be removed.

Although many organizations still fly traditional aerial photography where the output is acetate negatives and paper photographic prints, many are using the combination of aerial photography to capture the planimetric information, airborne GPS to track horizontal and vertical location during data capture, and LIDAR to capture the digital elevation information necessary for developing topographic data and the digital orthophotograph.

Dealing with the Images

Once processed into digital orthophotographs, the data are usually delivered in a set of files, most commonly using the tagged interchange file (TIF) format. This file format is standard in the graphic design and digital photography areas. If you are going to use these files in a GIS, there is the additional concern of adding locational information to the file. A TIF, by itself, is in a local coordinate system where the pixel at the upper left-hand corner of the image is at 0,0 and the pixel at the lower right-hand corner is at whatever the number of rows and columns are in the image, usually quite large. This raw image will not be properly placed behind your vector data without additional information. This is provided in an associated file, the TIF world file (.tfw). This file contains information on the projected coordinates of the upper left-hand corner and the number of real-world units covered by each pixel in the x (longitude) and y (latitude) direction, the cell resolution in the distance units of the projection (usually feet or meters). Other graphic formats such as .gif and .bmp, which are also widely used in graphics and digital photography and on the World Wide Web, do not support these world files and are not useful as backdrop formats.

The number of image tiles that will be produced from the photography for use in your GIS is a different decision from the number of photographs that must be taken. The orthophotographs can be digitally combined into tiles of any size for display. Unlike the individual photographs, the tiles will show a common size area and a standard distance in the x (longitude) and y (latitude) dimensions. Table 6.1 shows the relationship between cell size and the number of tiles. We selected 50 MB as an upper limit for a single tile because images larger than that draw slowly on normal desktop machines.

Table 6.1 Black and White 256-Gray-Scale Range

Cell Resolution (ft)	Square Tile Dimension (ft) Necessary to Keep Uncompressed Image = 50 MB	Area Shown on Tile (sq. miles)	Minimum Number of Tiles Needed to Cover 50 Sq. Miles	No. of CDs (670 MB) Needed to Store Tiles
3	12,247.4	5.4	9	1
1	7,071.1	1.8	28	2
0.5	3,535.5	0.4	112	8

If you will be using digital orthophotographs as backdrop data, there are implications that affect hardware selection. The graphic images are large and require
extra RAM in your computers. Redraw speed can be improved with higher-quality video cards as well, which keep a lot of the image processing right on the video
card rather than using the computer's core memory. Larger and higher-quality flat
monitors also enhance the use of backdrop photography.

Using file compression technologies, it is often possible to mosaic (combine) all
the tiles needed to cover your area of interest into a single image if your area of
interest is not too large. For example, a single image made up of mosaiced orthophotography (0.5 ft. cell resolution, 256 gray-scale levels) of a community of 24 square
miles requires 151 MB of disk space. Stored as TIF images it would take 2.6 gigabytes of storage space to cover 24 square miles. These images were compressed with
a proprietary compression program (MrSID, Lizard Tech) that can achieve compression ratios of 1:20 with very little loss in image quality; this example achieved a 1:17
compression factor. The DOQQs from the USGS are compressed with the Joint
Photographic Experts Group (JPEG) compression technology, which is relatively
lossy, which means that visual discrimination of features is reduced. Compression is
always a compromise between file size and information on the image, where the
uncompressed image always contains more information than the compressed image;
the trade-off is in the physical size of the image.

The ability to view compressed image files is present in most GISs, but the ability to compress them is a separate software license. If this is not an option or is too
expensive, it is not difficult to store the tiles in an image library or catalogue. With
some additional programming of your GIS, you can load and unload the needed
photo images tiles as you pan and zoom around the map. Either serious file compression or storing the images in a library or catalogue will be necessary to use
digital aerial photography in your GIS. You also need to make sure when you are
specifying your hardware needs that you account for sufficient disk storage space
(see Table 6.1) for all users of the system.

How Many Users of the Aerial Photography Will There Be and How Will We Serve Up the Necessary Images?

Image data, even when compressed to the maximum possible, are stored in very
large files. If you plan to store the images centrally and have users access them from
the remote server, you need to make sure that your network is capable of moving
files of 50 MB or larger quickly. Slow drawing speed frustrates all users. If this is a
problem, you should investigate the storage of multiple copies of the image data
on individual machines as another solution. Typically image data are acquired
infrequently, every 5 to 10 years, say, whereas the vector data draped over it may be
updated frequently, so a solution of storing many copies of the image data on the
desktop machines and the updated vector data on a server is a good choice. Each
desktop machine will have to have sufficient storage space and processing power
to quickly draw and redraw the images, though.

The advantages of using a digital orthophotograph behind your GIS data are obvious. Principally, it provides visual clues to where you actually are when you move around the digital map. When people can recognize the outlines of buildings, parking lots, and other geographic features they are familiar with, it produces a level of trust and a better understanding of the data. This is particularly true when the orthophotograph has been used to capture the base map data visible on the photographs. In that case the vector data lie exactly on top of the photograph. However, when you try to combine finely detailed image data with vector information obtained from maps at a much smaller scale, the overlap is not so good. Trust in the GIS data is not enhanced when it appears that roads (from the vector data) run right through houses (orthophotograph). Figure 6.3 shows a vector layer of roads (1994) from the 1:24,000 scale Digital Line Graphs of the USGS, which are used to create the 7.5-minute series topographic maps, draped on top of a 1:12,000 scale DOQQ (1995). It shows problems of updating; the subdivision road in the upper center was not captured in the 1994 vector data. It also demonstrates the problem of overlaying data produced at one scale on aerial imagery captured at a larger scale; the road running through the photograph is about 36 feet from where it appears on the photograph. This, of course, is within the published error standard of the data but still makes for an unsatisfactory combination.

Figure 6.3   Poor vector fit with DOQQ.

The decision to fly new aerial photography and incorporate it in your GIS is an important one, and you should certainly involve professionals in the process and not rely on the information presented in this section. The additional cost of creating digital orthophotographs from the negatives is small compared to the cost of flying and producing the original negatives, and the advantages gained from this background information are considerable.

How They Did It – Aerial Photography as Backdrop and Data Source: City of New York

In the early 1990s two staff people in information technology and city planning in New York City, Alan Leidner and Richard Steinberg, were lamenting the fact that the multiple agencies of New York City with mapping functions were using maps at different scales with different levels of accuracy, and this was seriously hampering coordination. They saw the need for an accurate base map for New York City, but it took 5 years and two mayoral administrations before the actual photography was flown. The city's Department of Environmental Protection, the Mayor's Office, and the Office of Emergency Management were the lead agencies that worked during the 5-year period to get the project underway, and the Departments of City Planning, Transportation, Police, Fire, Finance, and Design and Construction were all involved in the map design process.

The basic plan was to fly the city twice, once in conditions when the leaves were off and once when they were on. The plan was to capture detailed planimetric information from both sets of flights and produce color orthophotography for visual backdrop behind the planimetric information. The specifications called for accuracy at NMAS for 1 inch = 100 ft scale mapping, ±2 ft horizontally, and ±1 ft vertically. At this resolution, home plate at Yankee Stadium is visible, as are individual crosswalks and fire hydrants. They used airborne GPS and stereo pairs to produce the topographic information, the contours, needed.

With aerial photography those were the key spatial decisions to be made, but because the photography was going to be used to gather information such as curbs and building and block outlines, they had to decide which features they wished to capture from the photography in vector format. This almost always leads to heated discussions over what to include and what to exclude, and these decisions have significant cost implications. The total cost of this mapping was $5,000,000; $3,000,000 was the cost of capturing the planimetric data from the photography. Alleys and airshafts were included but not water towers and fire escapes. Awnings produced a lot of debate and one of the initiators, Alan Leidner, was quoted as saying, "Awnings! What do you do with them? And then driveways! When you capture them, what do you do with them?" (Gopnik 2000). There is a tendency to want to capture everything from the aerial photography, but going back to the question of what you will do with it keeps it focused.

The heads-up digitizing of the planimetric information was contracted out to a firm in Colorado and took a year to complete. Quality assurance and control are important parts of this process, and that is difficult to do at a distance (e.g., from Colorado). A local university, Hunter College, was hired to do the quality control and ground checking for the map, and this process produced over 30,000 changes to the map and took almost 2 years to complete,

There were geographic databases consisting of a comprehensive street name and address database, a tax lot and city block database, and a street centerline database that had been in place for some time in the City Planning Department. They have since registered that information directly to the NYCMap data. The city has a unique identification number for every building (BIN), and these numbers were attached to the building outlines captured from the photography at a cost of $1,000,000. These identification numbers are used in more than 20 city databases, which can now be tied to the building outline layer. The existing databases were rich and accurate in terms of their attribute information, and the NYCMap orthophotography and planimetric databases were rich and accurate in geographic information. The marrying of these databases has created what the city now calls a GIS Utility.

The actual process used by New York City to produce this GIS Utility was not revolutionary in any sense except for the scale of its development; New York City is a large, dense, and diverse place to undertake such a project. This project is also a good example of unexpected uses for digital orthophotography. During an outbreak of West Nile virus the Public Health Department used the photography to locate piles of automobile tires and places where water could accumulate in shallow pools to target their spraying program. The project and its photography have received a lot of media and public attention recently because of its use in the recovery during the World Trade Center disaster in September 2001. With that base map in place, derived from the photography, it was possible to quickly register the subsurface infrastructure layers to assist in the recovery effort. The success of the GIS and mapping efforts around this tragedy has been widely reported and could serve as an impetus for other communities to update their digital orthophotography and planimetric base.

Integrating Remotely Sensed Information with GIS

As Backdrop

The use of backdrop aerial photographs behind vector information in a GIS is principally for locational context. Seeing the outline of a building on a white background provides limited information to the viewer. When you can see the parking lot, the cars in it, the trees around the building, and walkways leading to it, the screen more closely resembles what you might see in the real world, although from a perspective you do not often get.

Remotely sensed information can be processed and used as a backdrop as well, and several companies are vigorously pursuing that segment of the market with panchromatic (black and white) and processed natural color products at resolutions of 1 to 15 m. The information is also used to develop layers of information useful in analysis or management tasks (e.g., land cover or change analysis). An aerial photograph brings a visual context, but remotely sensed information brings analytical content into your GIS. Technically, aerial photographs are remotely sensed information, as is MRI of the human body, but here we are referring to data gathered by sensors other than film in a camera (or its digital equivalent) from airplanes or satellite platforms.

The technology and science of remote sensing predates GIS and was the product principally of government activity driven first by resource management requirements and later for military needs. But now they have become inherently linked technologies due to the power of remote sensing to provide information to a GIS that is unobtainable in any other way (Star et al. 1997, xv). Only some GIS software systems have the capability of processing remotely sensed data, but all can display data that have already been processed. When you are working with remotely sensed information, most of the time any processing is to make the information useful for your purpose; typically the analytical tasks will take a relatively small percentage of the time involved. The technical problems of taking information from multiple sensors, over different time periods, and preparing it so it will overlay closely with the vector information in your GIS are considerable and, like aerial photography, require the skills of trained individuals.

There is no quick fix, no right-out-of-the-box way to integrate remotely sensed information unless you are willing to pay for professionals to do the processing. If you wish to do it yourself, there will be a significant learning curve. So, as in the previous section, our goal is not to make you experts in remote sensing in a few pages in a book but to expose you to the problems involved and the key decisions you need to make as you integrate these different but linked information sources. Also, the technology can be broadly split into passive sensors, which measure energy reflected by the earth's surface, and active sensors, which send out a microwave or radar pulse and measure the reflectance of that pulse off earth features. Our discussion is on passive sensors; data from active sensors are of little utility within a GIS with one exception, LIDAR, discussed earlier.

The advance of the technology of remotely sensed imagery has produced steady increases in spatial resolution, spectral resolution, and temporal frequency of coverage. Spatial resolution refers to the sensors internal field of view (IFOV), the size on the ground of each pixel recorded by the sensor (see Table 6.2). Spectral resolution involves the portions of the electromagnetic spectrum (EMS) the sensors can detect. Each band detects energy reflected from the earth within a discrete range of the spectrum (e.g., green 0.0 to .60 micrometers). Generally, remote satellite sensors cover visible light through the near and mid-infrared segments of the spectrum, sometimes moving into the thermal infrared segment with a larger cell size. Over the past 30 years there have been significant increases in the number of bands and narrowing of the sections they detect, which yields increased information about geographic features. Temporal resolution is how frequently the satellite returns to take information from the same area of the earth. Higher frequencies allow for closer monitoring, and because the system works best during periods of no cloud cover, a higher frequency will increase the likelihood of getting good imagery around the time you need it. If your need is for black and white or natural color background information, the key concern is spatial resolution: The smaller the cell size, the closer to aerial photography the imagery will appear. If you have analytical needs that depend on careful measurement of reflectance within certain segments of the electromagnetic spectrum, spectral resolution may be more important to you than spatial resolution. And if your need is for regular, high-frequency monitoring, temporal resolution will be the paramount issue. Selection of the right source or sources of remotely sensed imagery for inclusion in your GIS is principally a trade-off among these three concerns, the amount of money you have to spend, and how quickly you need the product.

The first sensors were the Landsat Multi-Spectral Scanners (1, 2, and 3) launched by the U.S. government in the 1970s. The scanner on these satellites had a spatial resolution, or pixel size, of 80 m and recorded reflected energy in four bands covering the green, red, and near-infrared portions of the electromagnetic spectrum. It repeated its coverage every 18 days. These early Landsat satellites are no longer in operation, but the archived data are still used for change analysis. The second generation of Landsat satellites, Thematic Mapper 4 and 5, carried the earlier sensors but also carried sensors that increased both the spatial resolution (to 30 m) and spectral resolution (seven bands plus a thermal infrared sensor). The increased resolution in both the spatial and spectral dimensions of these satellites significantly increased the usefulness of the information for smaller regions. A spatial resolution of 30 m is approximately 100 by 100 ft, and at that scale the information becomes helpful even at urban and suburban scales. The 80-m resolution of the first generation of Landsat satellites restricted its use to large areas and small scales. Thematic Mapper 5 was still sending information as of March 2002, but the status of satellites changes frequently.

A major advance in the technology occurred in 1986 with the launching of the SPOT satellite. This satellite, principally the operation of a French firm with Belgian and Swedish partners, was the first to collect panchromatic (black and white) imagery and did so with a significantly improved cell resolution of 10 m. For more than a decade this was the highest resolution satellite imagery available to the public for purchase. For the multispectral bands, they opted to use reduce the amount of the spectrum covered and reduce the cell size to 20 m. Additionally, the sensor can be pointed and can capture information off the vertical. This means that with repeat visits it is possible to capture the satellite equivalent of stereo pairs, which allows the calculation of a digital elevation model and some development of rudimentary contour lines. Accuracy in the vertical dimension is 7 to 11 m, which means it is not useful for large-scale (small area) work beyond 1:50,000. Five generations of this satellite have been launched, and data are still being

Table 6.2 Remote Sensing Systems for GIS

Sensor System	Spectral (Number of Bands and Wavelengths)	RESOLUTION		Time Period of Coverage	Status
		Spatial (meters)	Temporal (Return in Days)		
Landsat Multi-Spectral Scanner (MSS) 1, 2, and 3	4 bands (0.5 to 1.1 micrometers: green to near infrared, IR)	80	18	1972–1982	No longer functioning
Landsat 4, 5 Thematic Mapper	Same as 1, 2, 3 plus 7 TM bands (0.45 to 2.35 micrometers: blue to mid-IR) Thermal IR	30 120	16/8	1982–present	Functioning in 2002; provides 8-day coverage when used along with Landsat 7
SPOT 1, 2, and 4	3 bands (0.50 to 0.89 micrometers). SPOT 4 added a mid-IR band (1.58 to 1.75 micrometers)	20 multispectral 15 panchromatic/black and white (0.51 to 0.79 micrometers)		1986 to present	Operational
IRS (India)	4 bands (0.52 to 0.86 micrometers: green to near IR) 1 band (1.55 to 1.70 micrometers: mid-IR) Panchromatic (64, not 256, gray scale) Wide field sensor (red and near-IR)	23 multispectral 70 6 183	1995 to present		
Landsat 7, Enhanced Thematic Mapper	MSS 80 m; no longer on satellite Thematic Mapper; MSS bands the same Panchromatic (black and white, 0.5 to 0.90 micrometers) Thermal IR	30 15 60	16/8	1999–present	For large area monitoring Operational
IKONOS	4 bands (0.45 to 0.853 micrometers: blue to near IR)	4 multispectral 1 panchromatic	1 to 3	1999–present	Operational
SPOT 5	Same as SPOT 4	5.0 and 2.5 panchromatic 10 multispectral 23 multispectral			Operational
IRS (India)	4 bands (0.52 to 0.86 micrometers: green to near IR) 1 band (1.55 to 1.70 micrometers: mid-IR) Panchromatic (64, not 256, gray scale) Wide field sensor (red and near-IR)	70 6 183			Operational

received from four of them. There was a launch in April 2002 of SPOT 5, which will produce 2.5-m black and white imagery and 5-m multispectral information.

In the 1990s the U.S. government declassified much of the satellite scanner technology. This allowed several private firms to get into the business of building and launching low-earth orbit satellite providing high-resolution (up to 1 m) black and white imagery that approaches the appearance of a digital aerial photograph. The technical and financial concerns were sufficiently difficult that only one company (Space Imaging) did not put up a satellite (Ikonos) until 1999. The panchromatic resolution of these images is equivalent to the DOQQ from the USGS, but they have the advantage of timeliness. These satellites orbit closer to the earth and come around to the same place every 1 to 3 days. The production process for DOQQs is measured in years, whereas these products can be delivered in weeks. The appearance is close to photography and is usable down to a scale of 1:10,000. Although the sample images provided are often of dense, urban areas, when you zoom into these images, the 1-m resolution is too large for work at an urban scale (see Figure 6.4)

In addition to the satellites discussed and presented in Table 6.2, there are medium-resolution satellites (similar to Thematic Mapper) placed in orbit by India (IRS-LISS), Japan (JERS), and the European Space Agency (ERS). The Japanese satellite functioned from 1992 to 1998.

Figure 6.4   One-m Image of Deer Valley, UT, Feb. 17, 2002. Image was in natural color, reproduced here in black and white. *Source*: Space Imaging Corporation (used with permission).

Presently this high-resolution imagery is a viable substitute for black and white aerial photography at scales down to 1:10,000. If the current trends continue, within 10 years there should be panchromatic imagery available at 1-ft or less resolution and multispectral/color imagery at 1- to 2-m resolution. Because all these products can be geo-referenced to provide coordinates for the pixels and ortho-rectified to remove terrain-caused distortion, they will be a viable substitute for aerial photography at scales approaching 1:2000 or 1:1000. GIS operations that require higher levels of spatial accuracy, to 1 ft or less, will continue to rely on aerial photography gathered using GPS processed with digital photogrammetric tools. But for an increasing number of GIS installations, remotely sensed data are a real alternative and much less expensive than digital aerial photography.

Questions to Ask

There are some basic criteria that have to be met before you can successfully incorporate aerial photography or remotely sensed data as a backdrop:

- At a minimum, the data will need to be in the same map projection as your other GIS data.

- If your interest is in small areas, you will need to ortho-rectify the images to remove effects of terrain and camera tilts; this is a function that is beyond the capabilities of most, if not all, commercial GISs.

Some summary questions that need answers before planning to incorporate aerial photography or remotely sensed data as backdrop for your GIS are as follows:

- What is the smallest feature you need to recognize on the photograph? The smaller this is the smaller a pixel size you will need for your digital images.

- How current does the information have to be? If you need up-to-date information but only infrequently, your cost considerations are quite different than if you need it frequently.

- Is color or black and white imagery the best for the application and region? Natural color images require more storage space and computer display resources. Black and white imagery often has higher contrast, which makes it easier to distinguish features.

As Analytical Layers

The preceding discussion involved using the remotely sensed images as a visual backdrop, either in black and white or color, for GIS data. Some GIS operations (e.g., land management and forestry) benefit from use of remotely sensed data to perform analytical tasks such as determining forest or range health. This is where the worlds of GIS and remote sensing begin to part ways. Although both are considered part of the geographic information sciences, the issues of image analysis are sufficiently complex that we have chosen not to deal with them in this book.

Often what happens in this area of research is that an individual will become trained in analyzing the satellite imagery visually and will drift into GIS to drape vector information such as roads, political boundaries, and so on, over the processed remotely sensed information. Although GIS professionals will pick up enough understanding of aerial photography or remotely sensed information to bring a real-world context to the computer screen to make it look less like a map, remote sensing professionals will acquire enough knowledge of GIS to bring vector information to make the images look more like a map. They will work from different technologies to the same destination, one putting a background behind the important data and one draping context on top of that information. The two technologies are very complementary, and practitioners of both can benefit from knowledge of the other's practice.

ADDITIONAL READING

Gopnik, A. 2000. Street furniture. *The New Yorker* (Nov. 6): 54–57.

Lillesand, T. M., and R. W. Kiefer. 2000. *Remote Sensing and Image Interpretation, Fourth Edition.* John Wiley & Sons: New York.

Star, J. L, J. E. Estes, and K. C. McGwuirem eds. 1997. *Integration of Geographic Information Systems and Remote Sensing.* Cambridge University Press: Cambridge, UK.

INTERNET RESOURCES

Community Cartography, licensed reseller of NYCmap data:
comcarto.com/basemaps.html

How NYCMap was used during the World Trade Center emergency in September, 2001:
geoplace.com/gw/2002/0201/0201wtc.asp

Description of the efforts to create and register the underground infrastructure to the NYCMap, accelerated after September 11, 2001:
geoplace.com/gr/groundzero/deepinfra.asp

National Oceanographic and Atmospheric Administration "LIDAR Use in a GIS":
csc.noaa.gov/products/nchaz/htm/lidtut.htm

Implementation: Data Development and Conversion

Once the assessment and basic design stages of the GIS are complete, the next phase of an implementation is to develop a detailed implementation plan. That plan is the point at which you combine the information that was learned from the needs assessment with the knowledge from the design phase and develop a strategy to execute the implementation and achieve success throughout the life cycle of the implementation. There are a number of factors that need to be taken into account that ultimately result in an implementation plan that has all of the following components:

- System configuration and product architecture plan
- Data development and conversion plan
- Application development plan
- Staffing and management plan
- Implementation phasing plan

This chapter discusses the first two sections of this plan.

System Configuration and Product Architecture Plan

The system configuration and product architecture plan defines the fundamental components, or base foundation, that will make the system work. As the name infers, this includes two pieces: the system configuration and the product architecture. The system configuration is primarily composed of the design of the core back end that will make the system work. The back end of the system is closely related to the choice of the software solution to be used, but because most of the major software vendors have designed their software with an open architecture, they can usually support any back end. Two primary approaches are traditionally taken. The first is to use a file-structured system. This type of a system uses proprietary file formats to store both the graphic and attribute data. Examples of these types of formats include coverages (ESRI based) and mapinfo file format (MIF). The advantage of this type of an approach is that the software manages the data for the user, and a low-level technical knowledge is all that is needed to use, interact, and manage the system. The main weaknesses of this approach are that the system lacks security and can be cumbersome to work with, and in larger systems the data can be difficult for a user to find. This type of an approach only works well with very small systems such as for a single user or one to three users. Beyond this size, the second type of a system should be used.

The second type of a system is to create a centralized data repository in which all of the data in the system is stored and managed. Again, two different approaches can be taken to this type of system. For small systems a Microsoft Access back end can be used that will allow both the spatial and tabular data to be stored in one location. Metadata can then be created and attached to each data set to cure the problem for users of finding the right data set. From a security standpoint, an advanced user can program security into the front end of the system to restrict users from having read or write access rights to any individual data set. There are two main weaknesses to this approach. The first is the fact that this type of system cannot manage multiple users editing a single data set. Basically, if two users attempt to edit the same data set, the last one set that is saved is the one that is stored; the other user's changes are lost. This is only a problem when the organization needs to have multiple people editing the same data set. Simple routines can be written that can restrict a second user from editing the same data set or warn the user that another user is already editing it. Again, for smaller sites, this type of solution isn't a problem. The second weakness of this approach is that there is a limitation to the amount of data that can be effectively stored in Microsoft Access. The typical size of a file that can be effectively used in this format is a one that has less than 250,000 features. Once you go beyond this, performance begins to degrade, and unexplained problems start to happen on a more frequent basis.

With recent changes in the primary GIS software, products that use a more sophisticated back end are less of a problem. Most mainstream products support Microsoft Access, Microsoft SQL-Server, or Oracle as a back end. Both the latter two, more sophisticated back ends, solve the remaining weaknesses of the centralized data storage approach, but they also add the need for a much more sophisticated administrator for the system.

The product architecture component of the plan defines the architecture that will be used to disseminate data to the end users. There are two fundamental ways that this is approached. The first is through a distributed architecture in which each client installation has the applications installed on local computers and the data are stored in the central repository. All users typically access the data in this single central location, and maintenance of the data is also done centrally. The second type is client-server architecture. With this approach the applications are typically also stored and accessed from a central location, and little to no software is installed on the client side. The primary benefit of this type of the approach is with system administration. With a distributed system each time an update is made to an application, the administrator has to make this change to each client's station. Much of this can be done through automated software distribution mechanisms, but the purchase, setup, configuration, and implementation of these types of solutions can sometimes be costly and require a higher level of expertise.

Because in client-server implementation the applications are stored centrally, the updates and maintenance of the system are much easier and less time consuming. A typical example of a client server implementation is the Web-based solutions that are becoming so common today. The biggest weakness of the client-server approach is that there is usually a high level of cost and sophistication associated with the setup, configuration, and development of applications that will function properly with this method.

The latest developments with this type of approach are Web services. They are designed to provide a specific functionality based on sending the server a request and asking for a result in a formatted manner. For example, Microsoft's MapPoint Service is a geo-coding service where you send the service a request, provide the coordinates of where this address is located, and the service returns the coordinates of that location in the coordinate system and projection. Many large GIS implementations such as statewide systems, large cities and counties, and federal agencies are developing systems using Web services approaches.

Data Development and Conversion Plan

The data development and conversion plan details the data sets that will be developed, the sources that will be used to develop them, the methods that will be used for automation, the order that the data sets will be developed in, and the quality control procedures and measures that will be used to check each data set as it is completed. Most of the information for this plan has been acquired and documented through the needs assessment process, but at this stage of the implementation, the procedures that will be used must be more thoroughly documented. For each of the data sets from the needs assessment, a step-by-step process must be developed that details the operations and methods that must be used to automate the data set. The concept is to establish a procedure that walks you through a data set's creation and establishes the goal for each operation. As an example, we will discuss automating a parcel base from scratch using deeds, recorded surveys, and coordinate geometry to construct the line work.

The first step in this type of automation is to develop a detailed inventory of all of the appropriate source materials for each parcel's creation. The easiest way to complete this is to start with the best available mapping source, usually the existing tax maps, and a list of all of the parcels of land in the agency, which can usually be acquired from the assessor's database. Using the maps and the database, each available source map (subdivision plan, site plan, recorded survey) is reviewed starting at the most recently filed survey. Working backward in time, the parcels on the map and in the database are coded with a unique identifier representing the source map plan number. On the tax maps you simply write in the corresponding source map number, and in the database you create a new field and key in the source map number. Many times a single plan, such as a subdivision plan, can be used to automate many parcels. Once all of the maps have been inventoried, the next step is to identify the deeds that will be needed for the automation. Most assessors also track the current deed volume and page in their database to help with this process. Because we have just coded all of the source maps into the database, we can now query all of the parcels that do not have a source map. From this we can develop a list of all of the deeds that we need to automate for the remaining parcels. This list should be sorted in consecutive order so that the deeds can be collected in a sequential order, which is much more efficient than collecting them in a random order.

The next step in the process is to organize all of the source data. There are two ways that this is usually accomplished. The first is if you are going to work from hard-copy documents. A folder is created for each map sheet that will contain all of the source materials for this sheet. Sorting the parcels by map number creates a list of deeds and plans that is needed to automate each sheet, and a copy of each of these documents is placed in the folder. The more modern way to accomplish this is to use document imaging. With this process, rather than copying, sorting, and filing the deeds and plans into a folder, the source materials are scanned and put into a system, and each file is named by its corresponding source maps number or book and page number. The database and a basic imaging program can them be used for retrieval of the appropriate document when it is needed. In addition, these images will then be available after the automation for easy retrieval by the end users of the system.

Now that all of the source materials are inventoried, a technician can begin the actual automation process. During this step the technician would use a survey package to enter the line work using the bearings and distances, angles and distances, or the metes and bound descriptions from the source materials. Working systematically, the entire map sheet can be created, and then the technician can move on to the next adjacent map sheet until all of the parcels have been created.

The next step in the process is to tag each parcel with a unique ID so that the parcel can be linked to the assessor's database. Usually a unique parcel identification system such as map-block-lot or plat and lot exists that can be used for this linking. After the tagging is completed, an important quality-control check is done. Assessors track the deeded area of a parcel for the purpose of taxation. From the constructed parcel base the area of the automated parcel can be calculated and then compared to the assessed area. Variations in these two areas can determine parcels that have been automated incorrectly or parcels that have had conveyances after the source plans were created. An example of this is when one owner conveys the eastern 10 ft of their parcel to a neighbor. The two areas would be different, and this would signify that further research would need to be done on a parcel to compare its deeded area to the area from the parcel automated from a plan source.

All of these differences would need to be checked, and if such a conveyance has taken place, the parcel line would need to be adjusted for the conveyance.

The final step is to place the parcels over the base map that is used for the project and check to see if any problems such as parcel lines crossing buildings, garages, fences, paved areas, or other evidence of lines of ownership exist. In these cases the deeds again need to be checked for conveyances that may have taken place and corrections need to be made. It is important to consider one fact when this process is done. Many people who automate parcels assume that if there is a fence running between two houses, that must be the property line. This is not the fact. The only time that the fence is definitely the property line is when the fence is called for in the deed or shown on an accurate survey to be the line. It is actually quite common for people to construct fences or walls 3 to 5 ft off of their property line because they don't want to intrude on their neighbor's property, or they want to be able to run the lawn mower easily on the other side of the fence to mow their lawn. Hedgerows are another example of this. People also often plant hedgerows off of their property line to leave room to trim the hedge without having to go onto their neighbor's property.

The following example details the process that can be used to automate or convert utility features. Utility features are one of the most complex types of data sets that can be developed in a GIS. The reason for this is that they typically have been constructed over a long period of years, have been changed and maintained throughout these years, and are composed of features that are visible and those that are invisible, or under the ground. The biggest challenge with automating these layers is knowing where to go to find the data about these invisible features. The key to determining this is first understanding the sources you have available to you and the desired accuracy and completeness that you want in the data set and then developing an approach that will provide the results you want using what you have available to you.

With this said, the first and most important aspect of developing a utility layer is to determine the accuracy you want. Utility structures such as catch basins, manholes, and valves can be automated in a number of ways. Using different methods of GPS, they can be located to a distance of 3 to 5 ft or as close as the nearest tenth of a foot. The methods that can be used and the resulting accuracies are discussed later in this chapter. Once the accuracy is determined, an evaluation of source mapping that is available must be performed. In some cases highly accurate, as-built drawings may be available that can be used to digitize the features from. If these as-builts are accurate enough for the desired result, the maps can be used to digitize the features. Another concern that needs to be evaluated for this sources is whether or not they are available for all of the system that is being automated. Often the accuracy of these maps varies based on when they were created, which can result in varying accuracy of the final products. The final issue that must be evaluated is whether or not the sources contain all of the information that you want to have attributed to the features. For pipes the common characteristics that are desired are the size and material type, the age, and the condition of the pipe. The first three of these are often on the as-built, but the condition can often only be determined by a physical field visitation. For the structures the important characteristics are the material type and style of the structure, whether or not the structure has a sump, and the grate or cover type. The types can usually be determined from the sources, but again, often a field visitation is required to collect the other characteristics.

For the best resulting quality of data the best way to approach the project is to use a combination of the best sources and physical field collection. With this approach the first step is to use the source documents to create an inventory of all of the structures. Each source is first inventoried as to what area it covers, and the features on the source are then digitized to their approximate locations. Once these locations are created, electronic files can be created and uploaded to a survey data collector for GPS in the field. The approximate locations will provide the survey crews with an approximate location that helps them find the structure in the field, but they then locate the feature using either GPS or conventional survey techniques. As they collect the point, they code each survey location with the type of feature and any physical characteristic that can be seen from observing a structure. Figure 7.1 is an example of a page from a utility guidebook that can be prepared to help make sure that the crew codes each feature properly. This guidebook should be put together prior to sending the field crews out to assure proper coding, and a different figure for each type of feature should be created.

Once the feature has been located and the survey data have been reduced, the next step in the process is to collect the attributes of the features that cannot be seen, the below-ground attributes. Typically each structure is again visited, but this time the grate or cover is removed, and data are collected for the feature below the surface or in the structure. Measure downs are taken, where the crew measures and records the distance from where the survey point was taken to the invert of the pipe. The size and material type of each pipe is also determined and recorded, as well as the apparent condition. During this collection the crew can also collect data about the structure such as what is it made of, if there is a sump, its condition, and if the structure is full of debris or silt. Another important value that should be collected is a confidence factor with the data collected. Many times when collecting the details of a system like this, a crew will encounter a condition where it is difficult to ascertain exactly what the size of a pipe is, or there may be a significant amount of debris that makes it impossible to get a good measurement to the invert of the pipe. In these cases the data collected can be coded with a low confidence factor so that a user back in the office would know the data might be questionable.

Returning to the office, the data collected can now be reduced and attributed to the features to create a complete inventory. The only step left is to perform quality checks on the data to try to isolate any errors that may have been made in the collection process. GIS functions can be used to calculate whether pipes have negative slope, which may indicate that a bad measurement has been taken. Using the connectivity of the system, an analysis can be run that determines whether a pipe run starts at one size, say 24 inches, then changes to a smaller size, such as 18 inches, and then goes back to 24 inches. This could be an indication that the wrong size, 18 inches, was collected for that pipe because pipes rarely get smaller in a system.

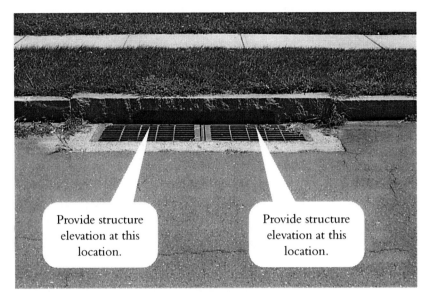

Code	13
Structure	Catch basin
Grate	Double Grate
Type	Type II
Curb	Granite
Inlet	Yes
Survey Locations (2)	Centerline Gutter

Figure 7.1 Utility guidebook example (catch basin).

These types of a detailed processes need to be laid out and documented for all data sets being automated so that a thorough understanding of the process is possible, the anticipated results can be predicted, and the creation process itself can be recorded in the metadata for the features. In the data development plan it is common to summarize this approach in the steps to follow and then to reference a more detailed technical specification that would contain the narrative and technical aspects of the layer development.

Capturing Digital Data

Depending on the type and nature of the data sets to be built into a GIS, there are many different techniques that can be used to create them. The primary methods that are used for construction on base features are photogrammetric data capture, digitizing, scanning and raster-to-vector conversion, and GPSs.

Photogrammetric Data Capture

Many organizations that use GIS rely on aerial photography as the principle source of GIS data. After the photographs have been turned into digital data, orthorectified, and placed in a map projection and coordinate system visible information on the images can be captured using specialized photogrammetric techniques. The capture of this visible information, called planimetric data, is done through specialized line-following software. This software allows technicians to place a cursor at the beginning of a feature, say the outline of a building, and the software will trace the remaining outline of the building. Through this process you can create a GIS record for any feature visible and distinguishable from other features on the photograph. This process requires specialized software and hardware and is usually done by the firm that does the aerial photography.

Digitizing

Digitizing is the next most common method used for data development projects. It is the process by which the operator converts a document that is in an analog form into digital form by tracing the features depicted on the analog form into the system using a combination of hardware, software, and operator assistance. There are three primary types of digitizing that are used today: head-down digitizing, heads-up digitizing, and automated raster-vector conversion.

Head-Down Digitizing

In head-down digitizing the operator attaches the source document to a digitizing tablet or table, registers the document to the base map using features shown on both the source document and in the land base, and then traces the features into the system. This technique was very common 5 to 10 years ago and was one of the most common ways to automate analog data into a system. The adjectives frequently attached to this data development process are tedious and time consuming, so it is not surprising that its use is decreasing. There are a number of factors that have led to this reduction. The first factor is the cost of the digitizing tables or tablets. A good-quality, large-format digitizing table costs $7,000 to $8,000 and is a very high maintenance piece of equipment. They often break down or develop dead spots that render them inoperable.

The second is improvements in computer hardware and reduction in costs for imaging software needed for other forms of data development. Five years ago the average price for a feature-rich imaging package was $10,000 to $15,000 for the software, and another $10,000 to $15,000 for the computer hardware to run this software. There were very sophisticated software products to learn and use properly and required computer experts that came at a fairly high rate of pay. Over time, the cost of both the software and hardware products have come done, thus decreasing the entrance cost to use the product, increasing the user base, and developing a broader user base.

Software procedures for heads-down digitizing are part of all GIS software packages, but many users will develop procedures using CAD software, which is easily brought into all GIS databases. The advantages to this approach to heads-down digitizing are the widespread availability of the software and the very large number of technicians trained in its use. Creating GIS layers of simple points, lines, and polygons is usually a much simpler process than creating the complex three-dimensional drawings CAD operators are used to. You must take care to ensure that the digitizing tablets are registered to the coordinate system you have selected for your database because the default is to return data based on the inches or centimeters of the tablet.

Heads-Up Digitizing

More improvements have come in heads-up digitizing, and this is what is commonly used. It is the process by which the source document is scanned into an electronic format, the scanned image is then registered to the base map, and the operator looks up at the screen and traces the features into the system rather than down at the digitizing table. The lack of the constant need to look down at the table, then up at the screen, and back again has huge efficiency gains and also eliminates the need for and problems with the digitizing tables.

Raster-to-Vector Conversion

The last method that is becoming very common is raster-to-vector conversion. In this process the source maps are scanned and geo-referenced to the base map, and then automated routines are used to trace out the features from the source map. The software used by the operator can perform two types of tracing. The first is semiautomated, where the operator selects a starting point, and the software then solves for the mathematical center of the object being traced and follows it until it comes to a conflict. The software then moves the operator to the conflict location, and the operator then chooses what direction to go in or what to do to resolve the conflict. The second method is fully automated conversion, where a set of parameters is configured and then the entire source map is automated all at once. The operator then performs cleanup using heads-up techniques to correct any issues with the automation. This method is also used to perform batch processing and automating many source maps that have the same properties. It is also becoming more common and is often used to automate source maps that are very consistent, such as parcel or topographic maps. This process is similar to the capture of planimetric data from digital orthophotos discussed above.

With all three techniques of analog (map) to digital (vector) conversion, there are issues of combining maps together. Often the area of interest is covered by a set of maps, and lines from a single feature will extend from one map to another. When you place the maps together you will often find gaps where the lines do not connect as they should. These gaps will be present in the digital versions as well, so you must specify standards and processes for matching up the edges of maps and how you will handle these inevitable discrepancies in the source data.

Optical Character Recognition

In addition to vector data, optical character recognition (OCR) conversion is being used to automate text into a system. Using OCR an operator can very quickly automate large amounts of text from a source map. The more advanced

OCR software packages allow the operator to distinguish size and style of text on the source map and selectively automate them or automate them onto different layers or into different database fields based on their properties.

In-House or Out-Source Data Development of Conversion

As already stated, data development or conversion for a project can be one of, if not the most, labor intensive and costly components of GIS development and implementation. Photogrammetric mapping projects require a great deal of highly specialized equipment and expertise including airplanes, expensive cameras (approximately one-half a million dollars), survey equipment, and analytical stereo plotters or softcopy workstations. It is cost prohibitive for agencies implementing a GIS to ever consider performing this type of development in-house, although a few attempts at this have been seen (none of which have been successful). Instead the implementer should consider being an active participant in the process.

The primary role for the average implementer to be involved is in the quality-control aspect of the project. This process really starts at the onset of the project with the development of a detailed specification for the base-mapping project. Most agencies look to specialized consultants who can assist the average implementer with development of this specification to assure that they will get the proper deliverables on a project such as this. Another option is to look to agencies similar to yours that have recently completed a project similar in nature. Their specification can often be reused, but it is important to find a project that is completed. The people who were involved in the project can be questioned about problems that may have come up during the project, and you can then see what changes can be made to the specification to eliminate the problem in your own project.

Selecting a Vendor

The next important aspect is to select a firm that specializes in this type of work, has a proven track record, is financially sound, and has adequate, experienced staffing to deliver the project on schedule. Experience with this type of project should never be held second to a lower cost that may be offered by a firm that has an unknown track record. If it sounds to good to be true, it is, and it is likely that the lower-priced firm just doesn't know what they are getting themselves in to. If they are reputable, they will at least deliver what they have promised, but it has been our experience that what you may save in initial cost, you will probably pay for in project delays. Don't forget that these types of projects are based on a flight and photography that is literally a snapshot in time, and the longer it takes to get the data, the older, and more out of date it will be. You may save money up front, but now you have to worry about updating your data.

Less-complicated development or conversion projects should be carefully thought about in the same way. In general, when projects are done in-house, they take longer to complete because of limited resources and time. The longer one waits for data for the system, the less useful it is. On the other hand, the initial cost to develop or convert data is typically higher when out-sourced then when kept in-house. This is true unless of course specialty equipment is needed or could be used to complete the automation process in a more efficient manner.

Another important factor to consider in deciding whether to complete the work in-house or to out-source is experience. If you have never converted the data you are about to convert, you have no experience, and it is quite likely that you will make some mistakes the first time. This will probably require you to back track and redo some, if not all, of the work. Another way to avoid this is to hire a consultant to help you develop the processes and procedures for completing the work, but don't actually have them complete the work. Most of the cost and effort of the project is the labor to actually complete the work, not the effort to plan it. The consultants can also provide you with labor estimates of the time it will take to complete the work using these methods, so you can make sure you have adequate resources to meet the time frames you desire for completion.

Costs are almost always one of the biggest concerns of developing a GIS. Much of the discussion thus far has been centered on doing it right the first time, which always leads to lower overall costs, but there are other ways to control cost as well. One is choosing and using the right labor source. Many local universities are now developing or have developed GIS programs, and many of them, especially the good ones, have developed internship programs for their students. This labor force can be one of the most cost-effective ways to develop GIS data sets, and using students also gives them valuable experience. Remember that these students are usually very inexperienced and that they need constant supervision to make sure they complete the work the right way and must have help resolving difficult issues they encounter and haven't seen before. The search for low-cost labor in GIS data conversion has led many companies to overseas locations, principally in South and Southeast Asia, where costs are lower, and there are even instances where corrections facilities are providing low-cost digitizing and scanning conversion services.

Another important factor in developing and converting data for a GIS is to make sure you do not convert too much information and that you choose the correct data format for the conversion. For example, let's say you are a water utility and you have 100,000 customers. For each customer you have a 3- by 5-inch service connection card that provides information about where the valve is to shut off or turn on the water service at the customer's location. On these cards you have all sorts of important information that tells you the type and size of the valve, the size of the main and the service, how the service connects to the building or buildings, the type of pipe is used, when it was installed, and so on. The cards are stored by the customer's account number in a series of file cabinets.

The typical problem that a person would be looking to solve with a GIS is to be able to respond to a service call such as: There is water coming out of the ground in my front yard. Can you come and fix it? What would commonly happen is that the utility staff person would look up the customer's account number, go to the file cabinet, find the index card, and then head to the location to fix the problem. GIS can solve this problem in two ways.

One way is to automate all of the 100,000 cards into the system and build a detailed network that shows where each of the service connections are. During the conversion process all of the data would be transferred from the cards into the digital model. Then the operator could put in an address, customer's name, or account number, and the system could bring you to that location and then provide you with a detailed map and/or report of all of the information contained on the index cards. This is a very useful tool, but it also very expensive to implement and maintain.

The second way would be to scan each of the index cards. Name the file with the same name as the account number, and then tag the appropriate geographic feature (valve, building, parcel of land) with the account number so that a link can be built, and the scanned version of the card can be easily retrieved. The application would then bring up the electronic version of the card in the same amount of time as it brings up the map and the report, but the total cost to implement would be only 10 to 20 percent of the other, more detailed method.

The main concern that one should have in deciding which type of approach to use is to understand the problem that needs to be solved and know what information is necessary to solve the problem and the most efficient way to make the information available to the end user. In the oreceding example this specific problem is solved by both approaches, but many other things can be done with the first approach that cannot be done with the second. For example, if you wanted to know how many valves of a certain type were located in your service area, you couldn't determine this with the second method without looking at each scanned index card.

Perform a Pilot Project

One of the most important aspects of performing a data development or data conversion project is to always perform a pilot project to test the methods, procedures, processes, and final deliverables of any automation project. The exact methods to be used should be followed for a subset of the data or a small representative area to make sure that the procedures will result in the desired level or accuracy, precision, and completeness. The pilot will also uncover issues that may not have been thought about prior to automation, and they can be addressed early on in the process and reduce the amount of time spent redoing work.

Once the pilot implementation project is complete, the procedures tested, and quality control done on the features, you are ready to go into full-scale data conversion whether in-house or with a vendor. All the steps described in this chapter are important for a successful conversion and should also be noted in some way in the metadata. GIS databases, particularly in large organizations, have a long life and usually outlive the careers of the people who created them. So careful documentation of the production process allows new users to understand and evaluate the information. The rule of thumb is that up to 80 percent of the cost of a GIS is the development of the data in the system, so careful implementation planning, documentation, and processing are much more important than almost any other step of the process.

ADDITIONAL READING

Drecker, D. 2001. *GIS Data Sources.* John Wiley & Sons: New York.

Montgomery, G., and H. C. Schuch. 1993. *Data Conversion Handbook.* GIS World Books: Fort Collins, Co.

INTERNET RESOURCES

National Center for Geographic Information and Analysis. 1998. The NCGIS GIS Core Curriculum for Technical Projects.
Unit 12. Planning a Digitizing Project:
`ncgia.ucsb.edu/cctp/units/unit12/12_f.html`

Unit 13. Digitizing Maps:
`ncgia.ucsb.edu/cctp/units/unit13/13_f.html`

Unit 16. Planning a Scanning Project:
`ncgia.ucsb.edu/cctp/units/unit16/16_f.html`

Unit 17. Scanning Maps:
`ncgia.ucsb.edu/cctp/units/unit17/17_f.html`

Digitizing tools:
`giscafe.com/Download/Digitizing_Tools/`

Methods for creating spatial data:
`gislounge.com/features/aa092700.shtml`

Implementation: Selecting Hardware and Software

S oftware should be selected based on the functionality it offers and the applications that have been identified from the needs assessment process. Your hardware selection will be based on the software you select, and the operating system will be based on the product's requirements and the standards that have been developed by your agency.

Software Considerations

There are many different software products on the market today produced by various vendors large and small. Selecting a software solution is somewhat dependent on the operating system and hardware platform you will be using, but in most cases the products that are available have standardized on the Windows platforms and require a 32-bit operating system. Initially you should evaluate the software solution independent of the hardware.

The nature of hardware and software technology is that it changes, typically at a very rapid rate. In recent years this has been particularly true. New advances in technology, new software development environments, and drastic changes in approach by many of the software vendors have lead to an increased amount of confusion and concerns by the intended users. Existing users that have been using a certain product line for a long time have been hesitant at moving to new versions, and new users have been unsure of what to select because of the hesitations by existing users. This is actually very common historically with software. Some users who jump in with both feet and try the newest version as soon as it is available are considered early adopters. They not only use products when they are brand new, but they often use products during alpha and beta testing periods. Alpha testing is the period in which software vendors are trying to decide which new functionality will be included in the next version of their product. They typically solicit feedback from existing users through user's conferences, focus groups, and mass mailings and try to decide what the masses are looking for. They also typically purchase copies of competing products and review these products for features that are not currently included in their own products. From this list of potential features and functionality the software vendors develop a product enhancement plan that details what they will be including. They then involve a limited set of alpha testers in the program. In this program the vendor typically begins designing and developing the new product enhancements for two reasons; the first is to determine what the total effort will be to add the feature or function to the product, and the second is to make sure that it understands what the users are looking for. Through this period the final decisions are made about what will actually be included in the product.

Beta testing is the name for the time period that a product goes through just prior to release. During this time there are often a large number of bugs with the product, and early in this stage it is unstable. The functionality that will be included that was ironed out during the alpha testing period is now locked in, and the features and functionality are in full development. Those who are involved in this period are typically the more sophisticated users who like to play with new software. They are asked to test each of the new functions and report on any of the problems or bugs that are encountered. The beta testing period itself usually goes through a series of stages that are usually numbered sequentially (i.e., beta 1, beta 2, and so on). As each sequential release of a beta product comes out, it usually becomes more stable and usable. Products that are in beta should never be used as the primary solution for any system; they should always be used in a separate testing environment. It is not uncommon that data that are used with a beta product can be corrupted by the software and could be rendered completely useless once corrupted. Be cautious about being in an alpha or beta program. Your involvement in the program can have a tremendous amount of influence on what comes out in a new product's release, but this involvement can be very time consuming and trying. Even some of the earliest of adopters have gotten frustrated with these programs.

The other type of user is typically referred to as a late adopter. This type of a user is more of a skeptic, a person who waits for others to work out the problems and then changes to the new technology when it isn't new any more, when it is proven. This is one of the first decisions you need to make about choosing a software product: Are you an early or late adopter? If you are a late adopter, don't choose a product when it has just been released; wait for the second or third release before you use it. Software releases are typically signified by a series of numbers. The standard for many years has been that when a product is first introduced, it is referred to as version 1.0. As bugs are worked out of it, and new releases or patches are introduced, the number to the right of the decimal place is indexed. In other words the next version would be 1.1, then 1.2, and so on. If you are a late adopter, you should probably wait until version 1.1 or 1.2 before you even attempt to upgrade to or use a product.

When a major release of a product is developed, the version number to the left of the decimal is indexed. So the second major version would be version 2.0. Unfortunately, sales and marketing folks have caught on to the fact that users often will not be as receptive to new products and will not purchase them until they are more mature. As a result they have started releasing initial releases of products that have higher version numbers than version 1.0. Be leery of this approach and always make sure you ask the vendor how many versions and releases of this product there have been. Another change that was first introduced by Microsoft was not to give a product a version number, but to name it based on the year it is released. The first time Microsoft did this was with the Windows 95 operating system. The one benefit of this was that it gave the end user an idea of how long the product has been around. Some GIS software vendors did this as well. For example, AutoDesk introduced AutoCAD MAP 2000.

The final practice that is worth pointing out is that products can often jump a number of releases. ESRI recently did this with their ArcView product line. ArcView was initial released at version 1, went to 2 and 3, and had a series of interim releases 3.1, 3.2, and 3.3. The new version of the product that was released was such a change that they released it as version 8. The theory behind this is that if you add up the total number of releases, full and interim, they would add up to eight versions. Realistically this is just a marketing ploy that makes a user feel like the product has been around longer. Other GIS vendors are are also using this type of a process. AutoDesk, which is known more widely as a CADD software vendor, but also plays fairly heavily in the GIS arena, was one of the first to do this. They started with a series of version in the 1s, then some in the 2s, 2.5, 2.6, and 2.7, and then jumped to version 9 of the software.

The important thing to recognize or question is whether or not the latest version is really a complete new product or a refinement of the previous version. Has it been reprogrammed from scratch, using a new development environment and new standards, and does it have a completely new look and feel? If it does, it will probably be much less stable for a longer period of time, but ultimately it will most likely be a good thing for the product and the users. These type of redesigning efforts take a tremendous investment by the software vendor, and they are ultimately a great advancement, but the early adopters of the product will have to deal with all of the typical issues of a new release. Once the products reach a later release, late adopters will be able to use the product without many concerns and take advantage of the investment of the software vendor.

One of the most important decisions that should be made is to select a product that has been proven in the marketplace and continues to have a clear development path. Avoid technology that is outdated or is on the bleeding edge and has not been proven, and be careful of marketing gimmicks versus actual facts about the product.

Evaluating Software

There are two primary factors that need to be evaluated when selecting a software solution: functionality and performance. Standards are also important. During the needs assessment process specific applications and the functionality required for those applications have been identified. During the software selection stage is when you take the information from this process and apply it to your selection.

Functionality

Functionality can simply be defined as the ability of the software product to do the things you need it to do in a straightforward manner. The first factor that should be considered is the graphical user interface (GUI). If those who are going to be using the product are generally novices with computer technology, it is critical that the product have a straightforward, easy to learn GUI. If they are more experienced users, the ease of the interface isn't necessarily as important as the functions that are available to the user. Most hard-core GIS users who have been around for a while do not really care about the GUI. Most of them grew up with the products before a GUI was even introduced. For these users the most important feature of the products was the command line.

Both ESRI and AutoDesk products were initially very command line driven. Typing in a series of commands on the command line performed all of the features and functions that could be performed by the product. One of the commands allowed the user to display the data or the results of the analysis on the screen in a graphic format. The next advancement that came was scripting languages. In the case of ESRI's ArcInfo product, the scripting language was AML.

For AutoDesk it is of LISP (List Processing) called AutoLISP. Basically these scripting languages allowed the users to store a series of commands that would be used over and over again, and rather than typing the commands over each time, they could call these scripts when needed.

This all changed with the creation of the Windows environment and the addition of the GUI. Most new users now choose the command they want from a button or pull-down menu and don't use the command line. The older users still use the command line to execute many functions, but they often use the GUI when it suits them. With the latest releases of some of the products the command line has been removed in its entirety. This has no impact on newer users, but as users become more experienced, this is likely to be a detriment to their performance.

Another factor that needs to be considered is the ability of the product to be customized using industry standard programming languages. Some products have little to no ability to be customized. They are designed to perform a specific function and perform it well, but that is all. If this is what the end user needs, just this functionality, this is not necessarily a limiting factor. The key is to understand its limitation before you implement it and make sure it does everything you need. The second level of products are those that can be customized using some type of an internal programming or scripting language. These products are one step above those that can't be customized. Typically these products have a customization language that has been developed by the software vendor. It is good that they can be customized, but the problem is that the only people who have experience and know the language are the actual users of the product; this includes the vendor itself and some consulting companies. If the product has a broad user base, this may not be that much of a limiting factor, but if it does not, it can be significant. The use of a standard programming language can open the doors to many potential developers for customization of a product.

The final level of customization is those products that can be customized using an industry standard programming language such as Visual Basic. When a product is designed in this manner, it provides for a much broader base of people who are knowledgeable about the environment and can customize it. The only issue is that industry standards change very rapidly, and it can be difficult to determine what is a standard and how long it will last. At the time of writing this book, the standard that is typically used for development of new application is COM technology. This will probably be around for a while, and the Visual Basic programming language seems to be the most common standard for customization. If you are looking forward, it is a pretty safe bet that staying on top of what Microsoft is doing will be a good choice. With this in mind it appears the their .NET technology is going to be the next wave. Many product vendors are already moving toward .NET technology and Web services solutions.

In summary, if the agency needs to develop specific customized applications, the software should have and use a programming language that allows the software to be modified or customized. If you are not going to customize your system, make sure you choose a solution that provides you with all of the functionality you need out of the box.

During the needs assessment process, the implementation strategy document that was developed included tables and matrices that defined what functionality is needed for the system. During the software selection process, the next step is to compare each product you are considering and determine whether the features and functions that are required are available in it. When you perform this analysis, you don't just want to see if the feature or functions are there; you also need to rate the software based on how easy it is to use. This can be done in two ways. The first is by having the vendor come in and demonstrate the product, performing the specific tasks you are looking for. The second way is to visit people who are already using this product and have them demonstrate it and provide you with their impression of its ease of use and its limitations. The second method is recommended, but it is also worth talking to the software vendor after you meet with users to discuss the concerns that have been raised. Often end users may know how to do something one way, but there is actually an easier way, and sometimes they may not think the product does it at all, but it actually does. The software vendor should be able to address all of these questions and concerns in their presentation, but don't get oversold by the sales pitch. No product does everything and many times one product does something very well, but does other things very poorly.

The primary difference among products today is with the way they handle raster and vector data. Most products have been designed to support either a raster- or a vector-based solution and only provide functionality for the other model as a secondary capability. ESRI products, ArcGIS, ArcInfo, and ArcView, were primarily designed to support vector data models, although they have some support of raster-based products within their core. In addition they can be extended through additional purchases of product extensions that provide excellent raster support and functionality. Similarly, AutoDesk products also primarily support vector-based models, but they can also be extended with the addition of a number of third-party add-ons to provide more robust raster support. Idrisi and ERDAS provide excellent functionality in the raster-based world, but they only provide a subset of the functionality of other products in the vector world.

Standards

As already stated, standards are an important aspect to selecting a software solution. The development and customization environments are two examples of where standards should be adhered to. Another standard that is important is whether or not the product complies with Windows standards. Microsoft has a series of standards that relate to a product's look and feel as it compares to a published set of Window's compliance standards. This is important because when a product is developed following these standards, certain functions are in the same place, and users who have already used another Windows-compliant product can learn it much quicker. For example, all products that meet this compliance standard have a File pull-down menu followed by an Edit pull-down menu starting from the left side of the screen. A user who has used any desktop software such as Microsoft Word or Excel knows where to find the standard functions such as Save As or Undo on these menus. If they were moved to, say, the third pull-down menu, the user would have to search for them, and the product would take longer to learn. There are many other aspects to the design of a product that make it compliant or not, but these are beyond the scope of this book. The important thing to remember is that compliance with mainstream standards can be crucial to a product's acceptance and adoption. If further information about this standard is required, just go to Microsoft's Web site.

Another important factor with respect to a software product and standards is the data format that can be used by the product. This not only includes the native file format of the software but also the different formats that the product can import and export data from and to. There are a number of standards and standard formats that are out there and supported, but the best ones are those that are developed by neutral trade organizations or by federal agencies. The two most prominent standards today are Open GIS Standards (OGS) and Spatial Data Transfer Standard (SDTS). OGS is an organization that is developing standards for developers to use as they design and engineer software. It is made up of representatives from all of the major software vendors. SDTS is a standard that has been created by the USGS for developing, documenting, and transferring data from one platform to another. Any product you are considering should be compliant with at least one of these two standards, and the software vendor should be an active participant in these groups.

Performance

The performance of any software product is dependent on two primary factors: the speed of the hardware it is running on and the way the product has been designed or engineered. If the product has been engineered well, it will be designed to take into account all of the resource available on the hardware on which it is running. If it has been engineered well, the next factor is what is actually available to it in the form or system resources (CPU speed, memory, disk space, etc.). GIS software is very complex and demands a large amount of system resources. The more complex the software and the functions it performs, the more resources it will need.

Performance of the overall GIS system will be significantly affected if you do not provide the appropriate level computer. Because computer technology is constantly changing, and software versions are also constantly changing, the best way to stay on top of these needs is to visit a software vendor's Web site and find the vendor's most current specifications to see what is needed. Also, if you are purchasing a new system, try not to go with the minimum specifications; these usually provide borderline performance. It is best to go with the recommended or preferred specification. Also make sure you look at the date when these specifications were last updated. If they are over 3 months old, call the software vendor and ask for a more current version. You can also get additional input from a consulting company or other users to find out what their experience has been. These recommendations will give you a more accurate idea of the type of configuration you will need and want.

Expandability of Products

The one factor that is a given with software is that it will change. The other factor that is also a given is that from the time you start with your implementation to the time you become and experienced user, your needs will also change. Another critical aspect to your selection of a software solution is its ability to expand with you. Expansion can be defined in a number of ways. Does the product have functionality that you will not use initially, but as your needs grow and your system becomes more mature, you will need then? Does the software have the ability to share data with other applications? As the system grows, does the software have the ability to communicate with other products through an application interface (API) or through other programming languages? In virtually all implementations one single product will not meet the needs of all the end users. This will mean that there will be multiple products in use that will need to share data and to communicate with each other.

Licensing Options

GIS software is not purchased; it is licensed. There is normally a one-time license fee with an on-going maintenance fee that provides you with the most current versions of the software as they are released. There are two primary types of licensing agreements that you have to choose from. The first is a per-CPU license. With this type of agreement an individual license fee is paid for each of the computers that the software will be installed on, and the license is tied to that computer. The second type of a license is what is called a concurrent, or floating, license. In this type of licensing a counter is typically installed on the server that tracks the number of simultaneous users using a specific product. Each time a user starts the application, a license is checked out until the maximum number of user licenses that have been purchased is checked. At this point the next user is told that all licenses are in use and will not be allowed to get in. This type of licensing is quite common and is usually the preferred method because fewer licenses can be purchased by the agency to serve more people.

Licensing is also becoming more complex. Because many products are now being developed using standards such as COM, the components that make up a software product can also be licensed. This can be very complex to both track and administer, but it can also have some tremendous benefits, primarily associated with the licensing costs. For example, ESRI's newest licensing mechanisms support floating, or concurrent, licenses; standalone, or CPU, licenses; and component licensing. Because their latest release of its products, the ArcGIS product lines, are developed and made out of all of the same components, it is possible to share the components of ArcView, ArcEditor, and ArcInfo. User can switch the product they are using to perform a task, and by design they can choose the lowest-level product that they need to perform the task. The problem with this is that it is a manual process. If a user is logged in using an ArcInfo license, but only needs functionality of ArcView, an ArcInfo license will be checked out unless the user manually switches to ArcView. A better model would be for ESRI to design the product to automatically check out the license based on the function being performed and the object being used rather than the software product chosen by the user. This type of licensing will be coming in future releases of many of the common products in use today.

How to Select Your Software
Evaluation Team

The first step in selecting software for an agency is to form an evaluation team that is made up of the ultimate end users of the system. This team should be made up of interested staff from departments involved in implementing GIS within the agency. These individuals need to be objective and not have predefined ideas of what system they want.

Prepare a Specification

The next step in the process is to prepare a detailed specification for the software. This should include all of the items discussed above, including the functionality that is required, the data that the software will need to be able to read, the preferred licensing scheme, the number of licenses that will be needed, and the installation and training services that you want to get the software installed and operational. Talk to agencies in your area that have already gone through this process and see if you can get a sample of the specification they used. Be careful to make sure that they are similar enough to your size so that the specification will fit well. Use these specifications as a template to work form, and customize all aspects of it to suit your needs. Once you have a draft of the revised specification developed, have an objective third party look at it. This can be another agency similar in size to yours (maybe one of the ones who gave a copy of theirs as a template to use) or a consultant helping you with the process.

The best way to get started with this process is to either contact other agencies who have recently developed similar specifications or hire a consultant who specializes in this type of work. Use the sample specifications as guides and ask the people who provided them what worked and what didn't work with the specifications. Adjust your specification accordingly and adjust the scope of what you are asking for to meet your needs. If you decide to work with a consultant, make sure the consultant has no vested interest in developing the specification. Preclude them from actually proposing on the software solution to assure their neutrality in the selection process.

Prepare a Formal Request

There are a number of different approaches to selecting the final solution you will acquire. A request for bid is a process by which you prepare a very detailed specification of what you want and request prices from potential providers. You then select the bidder who provides the lowest cost for the requested products. This process only works well if you are requesting very specific items such as three copies of a certain software product. Be very careful of exclusions or qualifications in a proposal. Profit margins on software solutions typically run 15 to 25 percent, and if someone's proposal varies more than 5 to 10 percent of all of the other bids, that vendor probably has forgotten something, and you may be heading for problems. If it looks too good to be true, it probably is.

The next type of process that can be used is a RFP. During this process a less-detailed request is put out that asks for a proposed approach, qualifications, fees, and a schedule to provide you with what you are looking for. These are typically reviewed based on a weighting system where points are given to the proposers based on all of the categories, and the highest score gets the contract. This process is best used when to scope of the service is still fairly specific, and you want to weed out firms that do not have any past experience or haven't performed well on previous projects and to select the best available combination of qualifications and cost.

The final process is a request for qualifications (RFQ) process. This process is virtually the same as the RFP process except that cost is not included. In this process it is felt that the best solution, the best products, and the most qualified vendor are more important than cost. Remember, a GIS implementation is typically long-range project, and first cost is never the last cost. Working with the right vendor with the best solution will ultimately give you the best and most cost-effective solution. It will also eliminate a lot of headaches for you.

Another important consideration is learning about potential consulting companies or vendors to work with on your implementation. You will want to invite qualified companies to propose on your project. The best source for this is to go to trade shows or GIS user group meetings and ask around. Again, try to stay objective. Don't get misled by flashy marketing ploys. Look for experts who have done what you are looking for and have quality reference sites. Talk to other agencies and get recommendations of companies they think are qualified to work with you, and don't be afraid to visit their offices and see their operations. Get to know the people that a consultant will be using on your project and make sure you understand their depth of expertise. Do they only have one person who knows how to do a specific task on your project? This could be a potential bottleneck.

Evaluating Proposals

Evaluating proposals should be done by the committee that was established at the onset of this process. Each committee member should receive copies of what was submitted, attend any formal presentations by the vendors, and have adequate time to review the submission, ask questions, and check with reference sites. It is important that the criteria by which these companies would be evaluated be established before the review of the submissions. Apples to oranges comparisons may happen if the specification and review process is not clear to all of the committee members. In addition, this entire process should be well documented in case a protest arises from one of the losers. If you have been specific and prepared a solid specification, the evaluation process should be straightforward. Some questions that should be asked when evaluating a submission are as follows. Has the vendor done the following?

- Proven it has experience and clear knowledge of the applications you desire and the functionality that is needed
- Described what it is going to do in a clear and understandable fashion
- Demonstrated that it has past experience providing these services
- Demonstrated that it has adequate, experienced staff to provide the services
- Proven that the project is important to it, and not just another project
- Shown that it can meet your desired time schedule
- Proposed a solution that fits with our available budget

Interviews and Benchmark Testing

Another way to review a product and a proposer's qualifications is to conduct formal interviews. At these interview the proposers should be given an opportunity to provide an overview of their companies, explain past projects they have worked on, explain why they are qualified to handle your project, introduce the key staff that will be working on the project so you know who you will be working with, and demonstrate and explain what their approach to your project is, and why you should chose them. These interviews should only take place with companies that you feel might be the right firm for you. Don't waste a lot of time interviewing firms that you have no intention in using and were weighted low in your initial review criteria. Sometimes you can learn from these other, less desirable firms, but more often they will say things in the interviews that will cause more confusion.

The final test that you may perform is what is known as a benchmark test. During this exercise you develop a series of problems, issues, or other functions that you would like the system to do, you provide the information and data needed to solve the problems or issues to the vendors, and you give them a specified time to solve the problems. Often you give them multiple problems so you can see how well the project manager organizes his or her team to solve the issues in an efficient manner.

During this process you will be evaluating the proposer on such things as the ability to do the following:

- Interact with staff in your agency
- Listen and break down a problem into its parts
- Solve technical problems
- Work together as a team
- Communicate effectively with you and with each other
- Present their findings, good or bad

Making the Final Selection

Once all of the reviews, interviews, and benchmark tests are completed, each committee member should be given time to evaluate each of the companies independently. Then, after the reviews are completed, the committee should come together and discuss their rating as a group. You will be surprised at the outcome. Many times there is a clear submitter who, hands down, is the most qualified, but sometimes there may be things about the team or personalities that are seen by one individual, but not others. It is very important to have adequate time to discuss the strengths and weakness of each potential solution or company. If there are still questions that need to be answered or concerns you need to address, don't be afraid to ask for additional information. Take the time to get all of the answers you need before making the final selection.

Once all of the questions are answered, and the selection is made, move quickly, get your contract in order, and get on with your project.

Hardware Concerns

When evaluating and discussing hardware, there are number of fundamental terms or concepts that you need to understand to be able to make a sensible decision. The following is a discussion of each of these concepts; however, your software selection determines the majority of the hardware requirements for you. Computer hardware is essentially composed of four basic components: the operating system (OS), the CPU, the hard drive or drives, and memory.

Operating Systems

An OS is the core software that makes all the other applications function. It is this program that tells the computer what to do, when to do it, and how to do it. Most people are already familiar with those that are in use today such as Microsoft's Windows products, Linux, and various kinds of the Unix operating system. Before any hardware purchases are made, it is important to decide what operating systems will be used in your agency. The plan should take into account the departments that will be using the computer system, the type of network being used (or being planned), what operating systems are currently being used, how large the database will be, and what kind of technical support skills you have access to (in-house or contractor). Higher-end operating systems like Unix have some tremendous advantages, but they come with a price and require a specialist to administer. Windows operating systems, on the other hand, have been designed to be more self-configuring and require less expertise to function. Of course, if you want it to function optimally, you will also need to have a specialist involved.

Central Processing Unit

The CPU is the part of the computer that actually does the calculations, or processes the instructions, that are sent to it by the operating system. The most common term that describes the processor's capabilities is the clock speed. This is stated in terms of megahertz (MHz) or more recently in gigahertz (GHz). The clock speed simply describes how many cycles per second the processor uses. The higher the clock speed the faster the processor. Another description of the processor's ability to process information is how many bits it can access at one time. Many of the new processors are 32-bit processors. This means that the CPU can access 32 bits of information during each of its cycles. There are still some older computers in use that are 16-bit processors, and now some of the computers that are available are 64-bit machines. Most of these run the Unix operating system.

Hard Drives

The hard drive, or hard disk, is the piece of the computer that stores information. The operating system you run, the applications that are used, and the data that are used in the applications are all stored on different types of hard drives. In working with your GIS you will quickly find out that GIS hogs disk space. No matter how much you plan into your system, you still need more. Now it is common to have 20, 40, or 80 gigabytes of hard drive on a single end-user machine and hundreds of gigabytes on your central server. Luckily the prices of hard drives have been coming down, so you can purchase these drives for next to nothing.

Memory

Memory on a computer comes in two primary forms. RAM is used as a temporary storage space by the operating system and by the applications that are running more efficiently to complete requests. The simple fact here is that the most applications run faster as the amount of memory is increased. This is true up to a certain point. For some application the more RAM you have, the better the performance is until you add so much that some other part of the computer, maybe the CPU, becomes the limiting factor. In these cases you can keep adding RAM, but the performance will begin to taper off after you reach a certain point. In addition some software applications can only use or address so much RAM. After a certain point they just ignore that it is there or don't recognize it. Again, most software vendors can provide you with data that indicates where this point is.

Networking Issues

The concept behind networking and the issues that must be considered when planning a network can be summarized into the following categories:

- The type of communication that will be used
- Local area network (LAN) versus wide area network (WAN) considerations
- The speed of the component hubs, switches, routers and network interface cards (NICs)
- The operating system and software that are running on the server
- Whether or not you will use a file server- or client server-based architecture
- Whether you will used centralized or decentralized servers
- Whether or not to use replication when there are multiple servers

Network Design and Speed Considerations

The network infrastructure and the network communications provide the fundamental mechanism that allows a centralized GIS to work. The types of network products that are chosen establish a stable and dependable environment by which the users of the system communicate with the central servers and each other. A variety of communication protocols exist today that support distributed applications, centralized applications, and both centralized and distributed databases. Before the advent of the Internet, network technology was a relatively static environment, whereas computer performance was increasing at an accelerating rate. Recent advances in communication technology to support the requirements of the Internet have caused a dramatic shift in network solutions. Some of these technologies as they pertain to a GIS implementation are discussed in the following sections.

The Desktop Environment

GIS applications require a significant amount of resources on the desktop and with respect to network performance. By its nature a GIS allows a user to have access to and analyze large amounts of data and present the results in a graphical format. Access to these data for real-time display and analysis puts large demands on network communications. Data must be transported across the network to where the desktop application is executed to display the information in an efficient manner.

Data that are used in a GIS system, whether they are in a spatial, attribute or an image format, are a collection of little pieces of information called bits. Each bit takes up the same space on hard disk when it is stored. For convenience, these small pieces can be organized and stored into bytes of information, with each byte containing 8 bits. Data can be moved from one location to another within packets that are groups of bytes. Data typically move across a network from one location to another through a physical wire, in fiber optic cables, or using microwave, radio wave, or satellite signals. Each type of network protocol that is in use has specific limits to the volume of data that can be moved based on the technology used to support the transmission.

Types of Networks

There are two basic types of networks that must be considered in as GIS: LANs and WANs. The volume of data that can be moved across a network is referred to as the network bandwidth and is typically measured in millions of bits (megabits) or billions of bits (gigabits) per second.

Wide Area Networks

WANs support communication between different physical locations. The technology that is used in a WAN typically supports much lower bandwidths than LAN environments, but the movement of data is possible over much longer distances in a WAN environment. The cost for WAN connections are relatively high compared to LAN connections because resources typically need to be leased from third-party providers such as phone companies to complete the necessary connections.

Local Area Networks

LANs function much the same way as a WAN, but there are some important differences. Typically LANs consist of a server with a network or network cards in it that are then attached to a hub through physical wiring. The hub then is connected to other hubs that are located around the building by a physical wire known as the backbone. From the remote hubs there are again wires that connect to each of the client computers using a NIC. In a LAN, the network bandwidth capacity is a function of the combination of the speed of the computer's NIC and the speed of each of the hubs in the system. Whichever of these is the slowest determines the ultimate speed of the network. Typical speeds are either 100 or 10 Mbps. In addition, newer technologies can now provide a throughput of 1 Gbps. As a minimum 100 Mbps should be used for all of the network components for any system that will be used for a GIS. The only exception to this would be if either a Web-based or terminal server solution is used, but even then the 100 Mbps is recommended.

There are two different types of hubs supporting most network configurations. A basic hub is the most common and works well for sharing data between LAN clients. Distributed client/server applications require much more bandwidth then these basic hubs can support. As a result a new type of hub was developed a few years ago, the switched hub. A switched hub can support much higher bandwidths and has intelligence built into it that allows it to better know what computers are attached to it and what information it needs to pass on to one of its computers versus what information it can just pass through to the next hub in the network. This type of technology and routers can drastically improve the performance of a network used for a GIS if configured properly.

Designing, sizing, and configuring a network is not something that should not be left up to a novice. Protocols, capacities, and technologies are still changing rather drastically in this area, and this is an important step that should be carried out by an expert in these areas. There are many companies that exclusively focus on this as a service because of its complexity. The concepts can be learned over time and with experience, but it is best if expert advice is sought out, or your implementation can seriously suffer.

Basic Client/Server Concepts

Applications move data over the network through proprietary client/server communication protocols. A combination of hardware and software located on the client and server computers define which communication format is used. Data that are transported are stored within what are called communication packets, which contain information required to transport the data from one location to the other and information about which computer the information should go to. This information is commonly referred to as its address.

A packet of information contains both the address of the source computer and the destination address. In addition, the packet contains information that allows the packet to be delivered across the network itself. The packet of information is referred to as the IP, which stands for Internet Protocol. IP packet size varies depending on the amount of data that are being sent, but the largest IP packet is around 65 kilobytes (KB). To support files that are larger than, this a series of packets are sent out that comprise the total piece of data that are being sent. A single data transfer can include several communications back and forth between the server and the client to complete an entire transfer.

Several client/server communication solutions are available to support network data transfer. Each includes a client and server component to support data delivery. The client process prepares the data for transmission, and the server process delivers the data to the application environment. Primary protocols used by GIS solutions include the following:

- *NFS (Unix) and SMB (Windows) protocol:* Typical used when a client needs to access data that are stored on the server's remote disk.

- *X.11 Windows protocol:* Typically used when only the display of the results of some type of information on the server (Unix based) needs to be transferred to the client.

- *ICA and RDP protocol:* Typically used when only the display of the results of an application on the server (Windows based) needs to be transmitted to the client.

- *HTTP protocol:* This is the standard protocol that is used by the Web.

Each of these protocols is used in the GIS environment depending on the application and the solution, the infrastructure that is in place, and the desired results and performance.

Client/Server Network Performance

The amount of data that need to be transferred and the network bandwidth are the two factors that affect the performance of a network. A typical desktop GIS application requires up to 1 MB of data to generate a new map display. This means that the system has to be able to efficiently transfer 1 MB of data across the network for each user to receive the requested results. Obviously, the more users who are on the system, the more data that are being transferred across the network.

Sufficient bandwidth is the key component to making this work. Another factor that affects the performance is the type of application that is used. For client/server-based applications, all of the data are transferred across the network to the client for processing. When selecting data from a file (coverage or shape file), the total file must be delivered to the client for processing. Data not required by the application are then rejected at the client location. This accounts for a considerable amount of network overhead that may not be necessary. Certain types of applications perform all of the calculations on the server and then only transfer the results of the query across the network. Typical applications that work this way are Oracle Spatial, ESRI's SDE, and Web-based applications. Rather than sending the whole data layer to the client, the first two of these types process the data on the server and select only the features that are needed by the client. The resulting subset of data is sent. This can drastically improve performance. Let's take for example a parcel data layer of 18,000 parcels. A typical layer like this can require approximately 35 to 40 MB. If a user were to window into a small area, say one block, the entire layer would be sent to the client, and the client would reject all but maybe a few thousand bytes. If one of these technologies were used, only the few thousand-byte file would be sent across the network, and performance would be drastically improved. Of course, you would need a much higher-end server in this system to process all of the requests from all of the clients using the system.

Using HTTP or Internet protocol, the same type of a process occurs. The request is received by the server and processed, and a graphic image of the results is created and sent across the network to the client. This is the most efficient transfer mechanism, but it can be limited to only certain type of functionality. Using some new protocols and languages such as Java, the vector data can also be streamed across the network to the client using a Web-based solution much like the other server technologies.

The Capacity of the Network

The total capacity of a network is a function of the size of the data that need to be transferred, the network bandwidth, and the total number of concurrent users. Only one client packet of information can be transmitted over a shared network segment at a time. Multiple transmissions on the same segment will result in what are called collisions, which will require a retransmission to complete the data's delivery. It is important to properly size a network to minimize the number of collisions that occur. This means that the network has to be sized so that it has the capacity to deliver the maximum amount of data that will be required across any segment during peak usage times.

Sizing a Network

Standard published guidelines are used for configuring network communication environments. These standards are application specific and based on typical user environment needs. Most software vendors produce technical specifications or white papers that provide direction about how to properly size your system based on size of data sets, the selected software solution, the network communication protocol, and the number of concurrent users.

Replicated Data

Another important consideration when designing a system is to consider whether or not data should be replicated within it. Replication is the process by which a master copy of the database is created, and replicas of it are then created, usually on remote or distributed servers. There are a number of reasons why you would do this. The first is for backup. Having a replicated copy of the entire system allows you to have a system to switch over to in the case of a catastrophic failure of the master database or server. In an enterprise solution, with so many people depending on the system, a replicated version would provide you with an efficient means of keeping the system operating. Another reason for a replicated copy is for data maintenance procedures. Often systems are designed with a primary server that acts as the distribution mechanism to the end users. Meanwhile, a replicated copy of the system provides a working copy where edits and maintenance are done on the data, and when changes are ready for distribution, the two systems are synchronized.

Functionally, the way this technology works is that a master database is created on one system or server. Replicated copies of this database are located on another server or servers that are known as replica sets. Based on configuration parameters that can be set in the relational database software that is being used, the replica sets synchronize themselves on a regular frequency. Synchronization is usually done when the users of the system are not there because a significant amount of network traffic is created by the volumes of data that are sent back and forth between the two servers.

A common issue that comes up in the replication process is the resolution of conflicts. A conflict is when a piece of data is edited in both the master database and the replica set, but they are edited to be different values. Rules can be set in the software that define how this is handled. One way of handling it is to define one of the databases as the master and the other as the slave. For example, maybe one of your departments is physically connected to a replica set server because that server is located in the building in which they work. Their edits are posted to the tables they have security to edit on this server, and they are the owner of these data sets. Another user mistakenly edits the same table and records in the master because they have the rights to, but they are not truly supposed to. For those tables the system can be configured so that the replica set is the master and changes that are made on this system are saved and the others are discarded. Another way that this can be handled is that when conflicts are detected, they are flagged and stored in a temporary location until a user reviews the conflict and commits the correct change.

Replication is an important consideration for a GIS system, especially an enterprise system. If the physical resources exist to use this technology, it should always be used. Even on small systems, using MS-Access as the back end, replication is possible and should be used, although it doesn't have as many options as a MS-SQL server or an Oracle database.

ADDITIONAL READING

Korte, G. 2001. *The GIS Book,* 5th ed. Chapter 14. Four Leading GIS Vendor's Products in Review. OnWord Press: Clifton Park, NY.

INTERNET RESOURCES

GIS Software Survey (2002) — Point of Beginning:
pobonline.com/FILES/HTML/PDF/0602survey.pdf

Some free GIS software:
gis.com/software/free_software.html

Designing the Organization for GIS

There are multiple design concerns when you are beginning the implementation process of a multiuser, complex enterprise GIS. In addition to the very technical (and important) design concerns around the spatial and attribute data, there is a management design concern: How this new system is going to fit into the existing reporting and budget structure of the organization. When you are implementing something as complex and expensive as a GIS, where it will be housed, how it will be budgeted for, and other management questions will arise early in the process. To some extent they are quite separate questions from the more technical ones such as What coordinate system should we use? and Should this particular field in this particular table be a text or numeric field?, but sometimes the answers to the former questions will lead you to a good decision about where control of the data should reside. The central organizational design issues are control and accountability. Both of these issues are wrapped up in the question of ownership of geographic information.

Ownership of Geographic Information

There are two ways to look at ownership: within and outside the organization. Ownership of the data inside the organization is something the organization controls; ownership when viewed from the outside the organization can be more complex. Inside the organization, in your design process, you need to make decisions about control and accountability for each layer, set of features in your database, and table in your database. These issues will arise early as it becomes clear to different units of the organization that data they previously regarded as their domain are going to be more easily shared and seen by other users. It is no secret that control of information in an organization is about power within the organization, so the introduction of a new way or organizing and sharing that information—the new GIS—will disrupt the existing lines of power within the organization. And when the questions about whose budget will be supporting the GIS arise, and they will early in the discussions, the power issues become even more important. While some design concerns are technical and best left to database experts, hired consultants, and so on, organizational design concerns directly affect how power is distributed within the organization, so they take the careful attention of reasonably experienced and high level people within the organization.

User Roles

If the GIS is going to support the activities of multiple units within the organization, a basic assumption of this book, these units all have a stake in how the ownership is arranged. Sometimes, in state and local government, there are legal restrictions on who may or may not make certain modifications to the data, and these certainly must be considered. But each layer or set of features, if you have adopted a database rather than layer model for your GIS, needs to have units identified that will be responsible for the spatial and attribute data. From an organizational perspective, it is the units, and roles individuals have within them, that are important, not the actual people themselves. In a business there may be a sales manager and because that role has certain responsibilities, the sales manager will have access to certain data in the system, regardless of who occupies that position.

People interact with relational databases through sets of defined roles and privileges. Roles and people are not the same type of thing because one person may, at different times, have several roles. The role describes your overall relationship with the database. You may just be a viewer who needs to see information, or your role may include responsibilities for modifying certain information. With a GIS and its two types of information, spatial and attribute, there are roles where the ability to modify attribute information is important, but some other role may be responsible for modifying the spatial information. Within a role is the notion of privileges. A viewer role may carry privileges only to view certain GIS layers or feature data sets and possibly only certain attributes within those data sets. Although there will probably be many roles in your final GIS installation, three basic classes of roles are easy to identify:

- *Viewer.* In this role the user has privileges to view only certain spatial and attribute information. This role usually requires some customization of the interface to simplify its appearance, and there may be custom applications that produce routine output. An example would be a clerk who produces maps and tabular output of property assessments or someone who regularly produces mapped output with standard symbolization over customized geographic areas for use in the field. Access to the public, whether through the Web or some other means, would be another example of a viewer role.

- *Modify Spatial Information.* This role entails addition, deletion, and editing of the points, lines, and polygons in the various layers of your GIS database. These are specialized skills, and this role will fall on a relatively small number of people. If a large staff is required to fill these roles, they will often make up a unit of the organization by themselves. This is common in GIS utility installations, for instance, where the database undergoes such constant revision. Databases that do not change as frequently will have fewer people who have this role.

- *Modify attribute information.* This role might be combined with the previous role and carried out by an individual, or people may have only this role and no other. It may make sense for the same unit or person who edited the spatial information to update the attribute information. A police officer who is completing an incident report is the logical person to both locate the incident geographically and to create and populate it with the necessary feature in the data set that will represent that incident. There are communication systems now that allow this kind of database modification directly from the police car. Given the legal requirements for reporting and tracking crime and arrests, you would want only trained people in this role who knew the meaning of all the various codes, what fields are required, which are optional, and so on. In other cases it might be more appropriate to assign the role of modifying attribute information to someone other than the unit that modified the spatial information. An engineering draftsperson in one department might be the right person to update the spatial information on a layer of water mains, but perhaps the supervisor who oversaw the installation of the main is the correct person to document the material, size, and other attributes about the installation.

- *Decision-making.* Some users will be responsible for making decisions based on information in the database. Some of these decisions may appear small and routine but are spatial decisions nonetheless. The decision to dispatch a fire truck to the scene of a possible fire is a spatial decision, as is a decision of what value to place on a land parcel. Decision making is a distinct role and may not involve a viewing role at all. There are many examples where decision makers only interact with the GIS through mapped and tabular output. If you wish to consider these roles as a hierarchy, this role is at the top, and units and individuals with this role need to be closely involved in the design process so that they get the required information to make the decisions.

Table 9.1 shows a simple example that might arise in local government involving these roles. Look across the rows to see what reasons different units might have for interacting with the data. The mayor, as important as he or she might be, should not have any authority to change land parcel information in the database. That would clearly be an abuse of power. But the mayor may want to have access to the data for any number of reasons, so there no reason not to make that player in the system a viewer of all layers (this only represents a subset of the layers or sets of features that might be involved).

Table 9.1 GIS Roles

Unit	Parcel Layer	Street Centerline (911)	Buildings	Crimes
Office of the Mayor	V	V	V	V
Office of the Assessor	MA,MD		MA – assessment	
Police Department	V	MD	MD	MA/MS
Fire Department	V	MD	MA – hazardous substances, evacuation assistance	
Planning Department	V,MD	V	V	
Engineering Department	MS	MS/MA – addition of newstreets and address ranges	MS – new and demolished buildings	
Public Works Department	V	MA – pavement management		
Building Department	V	V	MA – building permits	

Key: V = view only; MS = view and modify spatial data; MA = view and modify attribute data; MD = view and make decisions based on geographic information.

Say the decision was made to make the engineering department, because it has years of experience in CAD, the owner of most of the spatial information. When a new subdivision is accepted, it would be the responsibility of that department to make sure that the parcel lines are correctly entered and that correct addresses are linked to the street centerline layer for 911 emergency purposes. But because the police and fire departments make decisions about those addresses (i.e., dispatch emergency services), it is critical that the procedures the engineering department uses meet the needs of the emergency service users. Likewise, as buildings are added or demolished, someone must see that the correct polygons are added or deleted from the building layer, but if there are hazardous materials in the building or frail elderly citizens who would need evacuation in case of fire, the responsibility for managing that attribute information about the building should rest with the fire department. Or perhaps you would prefer that the engineering department perform those modifications as well, and if that is the case, you must have a

regular procedure for communication between the fire and engineering departments about the modification of that information.

The police department is certainly the only unit you would want to be able to
add, delete, or modify information in the layer representing crimes (e.g., a point
layer of arrests and police calls); the mayor may wish to view them regularly but
would have no other privileges. A well-designed GIS database might even block
certain fields within that particular table from even being viewed (e.g., the name
field in an arrest layer). The mayor may wish to be able to see yesterday's arrests
and their geographic distribution, but should the major be able to label that map
with the names of the individuals arrested? Within the police department itself
there may be specific roles and access assigned; not every patrol officer should be
able to edit other officer's data, and the chief may have access roles that are very
similar to the mayor's — viewing but no editing. Looking across the rows of a table
like this will give you a good idea of how GIS will be integrated into the working
of the unit. A unit with only viewing needs will have simple software and hardware needs. Looking down the columns you should see only one unit responsible
for modifying spatial data on a layer or set of features. That unit will require software or licenses for software and perhaps specialized hardware such as a digitizing
tablet to modify spatial data. There may be more than one unit responsible for
modifying attribute data, however, depending on the fields in the tables as they are
finally designed. One department might best enter some information over
another. In this example when a new land parcel is created, the field that contains
the address for that parcel should be the responsibility of the engineering department because it will also be represented in the address ranges of the street centerline file for 911 purposes. Generally, it is an excellent idea when building GISs that
will be used for delivering services to addresses to have one unit of the organization responsible for the main address file. This avoids lots of confusion about
addresses that look different but are the same (e.g., 124 N. Fourth St. and 124
NORTH FOURTH STREET).

The information necessary to create a table like Table 9.1 should come out of
the needs assessment process. Completion of such a table may be a two-step
process where you make the initial assignments of ownership and privileges as
soon as the layers have been defined and further refine the table after the detailed
tables and their attributes have been designed. This is because data viewers and
modifiers may use the same table, but certain fields will be locked out to certain
viewers, and you may not know what the fields are until later in the design
process. So the viewers in the assessor's office may be able to see all the several
dozen attribute fields it make take to create an assessment of a property, but other
viewers of this data may only need to have a limited set of information visible.
Creating those different views into GIS data tables is an important implementation step, but you identify the need during the design process and implement the
role after completion of the database. Roles are not difficult to define and assign in
RDBMSs, and you may discover new roles you wish to assign long after implementation is complete.

Staffing the Design and Implementation Process

In this section we are going to deal with staffing the design and implementation process; staffing the GIS after implementation is discussed in a later section. The key decision here is how much planning and design you want to do with in-house staff and how much you will contract for with consultants. For most complex, multiuser enterprise GIS designs, there will be some of both. It is rare for an organization that does not already have a GIS to have staff with the necessary skills. This is true even if one or more of the front-line units in the organization already has a GIS that supports its functions. That unit will be important in the design process for the organization because the new enterprise GIS has to meet their needs as well as others, but their application may be so focused and specialized that their experience is not as helpful as they might think. There may also be units in the organization that purchased a GIS, developed a small amount of data, and then never did very much with it. That unit is certainly not likely to have staff with the necessary skills.

How much and what part of the design and planning work you keep in-house and how much you farm out to consultants depends on two factors: the skill and interest levels of your existing staff and the time they have to devote to this process in addition to their existing duties. But even if you plan to contract out for the entire design and implementation process and pay consultants to do the work right up to installation and training, you will need a committee of your existing staff and managers to oversee the process. There will always be staff input and time involved in this stage, and there needs to be a structure for it. That structure is almost always a GIS coordinating committee, especially where the organization is in the service state, discussed above. Somers (1995, 1997) recommends a three-tiered committee dealing with policy, technical issues, and users, but this could also be handled as a subcommittee. This committee will have to go through all the formal and informal steps of formation, statement of mission, settling who will be in charge of what, and communications within and outside the committee that any *ad hoc* or permanent committee that cuts across units in the organization would experience. It will have at least the following functions:

- Overseeing the needs assessment and requirements analysis process.
- Developing and reviewing any RFPs that goes out to consultants. Sometimes in large projects a two-step process of a RFQ, which yields a reduced set of consultants who actually prepare the RFP, may be helpful. Although a GIS changes rapidly, the questions one should ask of GIS consultants do not. There is an excellent set of questions in Forrest, et al. (1990) to effectively evaluate GIS consultants. If you choose to have a consulting firm conduct the needs assessment and requirements analysis, that firm will be in an inside position when it comes to bidding for later, and more lucrative, implementation contracts, so careful selection of the various first sets of consultants you choose is important. Also, it is increasingly common for consulting GIS firms to have formal business

relationships with particular GIS software companies, which can sometimes affect their software evaluation.

- Reviewing and making recommendations on the proposals that come back. Selecting consultants is something most organizations have had some experience with, and selecting a GIS consultant is not much different.

- For any part of the process that occurs in-house, the committee will direct and oversee the overall staff input to the design and planning. For this function they need enough senior management that other staff feels compelled to participate. This coordination role is perhaps the most important function for this committee. Through their actions they can energize the entire organization around GIS, or they can make the process so difficult that the rest of the organization begins to see GIS as something to be resisted and avoided.

- Communicate regularly upward to management and downward to the rest of the organization what is happening with the design and implementation process. Some implementation plans have actually prepared regular newsletters on the progress of implementation. However the committee chooses to let people know that there is progress, it should not operate in an information vacuum. Let the entire organization know what is being done and how it is proceeding.

The committee may have other functions assigned by management in addition to those listed. In the formation of any committee in any organization, the decision of who should be on the committee (and sometimes what is equally important, who should not) is key. It is almost impossible to lay out guidelines for the structure of this kind of committee, but here are some suggestions:

- Keep it small but not too small. You should keep subcommittees to a minimum, but you may have to have them. Appoint a small number of people to interact with any hired consultants to simplify their lives a little. Regular attendance at the meetings is important, so the selected people need to have the right workload or be assured that they will be able to participate. Serving on a committee like this is always in addition to existing full-time duties, so assurances of this sort are necessary to get good involvement.

- Include skeptics as well as committed people. The conversion of skeptical users is key to successful implementation, and having them on the committee allows the committed ones to develop their skills converting the skeptical.

- When selecting members of the committee from technically oriented departments, choose those who have demonstrated ability to interact successfully with other units in the organization. Every technical unit has people who are very good at what they do but should be left alone in their unit to do it. These are not good people for a committee like this.

- Have a definite champion. The word *champion* has been used in descriptions of information systems innovation for at least 20 years (Curley and Gremillion 1983) and appears all over the dispersed literature on GIS implementation. A champion is an individual who is almost a missionary about spreading the news of GIS but is also articulate and carries some clout inside the organization. A champion is a practiced salesperson and knows how the organization works and how to get it to change in desired directions. Although a draftsperson in a public works department may be a vigorous proponent of GIS and very articulate, he or she is not likely to be the person who will be able to convince skeptical midlevel managers to get involved in the process. The champion need not be the chair of this committee but needs to be there. Often there will be multiple people who can play this role in different arenas.

- Include representatives who are current front-line users of geographic information. A committee completely composed of managers is likely to design a system that may look good on paper but doesn't work well. Users can be invaluable when it comes time to prioritizing existing applications for automation. Often, some of the functions a consultant or technical person may want to automate are functions that are being done very well with the current manual methods. All the GIS will do is force the staff person to learn a new, and not necessarily better, way to do the same task. People who actually do the job usually have a better idea of whether GIS automation is worthwhile than their managers might.

- Provide clear timelines and checkpoints for the committee's work. A committee without a deadline is a dangerous thing.

- Keep the meetings regular but not too frequent.

As anyone with experience in complex organization knows, committees are strange things. Some work very well, some fizzle out, and some are so dysfunctional that they almost never meet. Others were established at a time of need and never get disbanded and have a life of their own. They work only if the people on them, or at least most of them, have a reasonably clear idea of where they are going or can develop one quickly, work well with others, and follow up on what they said they would do. Running a good interdepartmental committee is something good managers should be able to do, so our suggestion would be that a manager should lead it. In an organization in the strategic state (the next section explains states) this manager may be high-level, but in a service state a logical choice would be a manager from one of the units that already have some GIS capabilities, if there are any. It is usually a bad idea to allow the committee to choose its own leader.

Where to Put the GIS

The decision about where to locate the various elements of the GIS you are going to construct and use is critical. There are many options available, and the culture of the organization will determine to a great degree which option you select and even the process you use to decide where to put these things. Those that do the basic work of the organization feel that they are already working at a full-time job and now some will be asked to do more. They are wondering, and rightfully so, if they are going to have the time and skills to do this new work. Managers are concerned about the same thing for their employees and themselves and want to know from whose budgets the resources will come for this expensive and time-consuming activity. And top-level management should be worried about the overall costs and benefits of a GIS and whether it is going to move the organization forward. These concerns come up very quickly as an organization works out where the GIS, or the pieces of it, will be located in the organization and what pieces of it you will contract out to consultants after implementation.

Huxhold and Levinsohn (1995) refer to a useful research project (Urban Information Systems, or URBIS) at the University of California. One major finding of this project (Kraemer et al., 1989) was to classify organizations into three different states:

- *Skill state.* The information systems department controls information technology and the distribution and use of software and hardware. If an organization is in this state, GIS is likely to be centralized in the information systems or technology department, and other users may see that it as too technical and somewhat isolated from their daily work.

- *Strategic state.* In this kind of organization information technology decisions are centralized at high decision-making levels, and the principal goal of information technology is to support the work of top-level management. In that type of organization GIS may be centralized in a unit that reports directly to top management, and other potential users may feel somewhat left out of the process and benefits.

- *Service state.* A condition that has multiple units that provide different types of services to varied client groups. They all have very different needs for a GIS. Implementation of GIS in an organization in either the skill or strategic set is a simpler process than if the organization is in the service state, as most are. Our discussion of where GIS should fit into your organization presumes that your organization is in the service state.

- *Mixed state.* This is an organization that cannot quite fit into any of the previous three categories. Parts may work like a service state organization and others like the strategic state, and the IT/IS department may feel that it controls a skill state organization. Huxhold and Levinsohn (1995) make an interesting conclusion for an organization in this state: hold off implementing an enterprise GIS until they can figure out what kind of organization they are.

Generally, there are three type of locations and associated advantages and disadvantages for an enterprise GIS within an organization, as shown in Table 9.2.

Table 9.2 Potential Locations

	Advantages	Disadvantages
Direct, front-line working units.	Direct link with operational needs and existing budgets.	Coordination is difficult, no authority of GIS units over other units, possible weak budget position, ownership is concentrated; other units have no stake, lack of visibility.
High-level executive placement.	High visibility, strong authority.	Perception of distance from operational needs, potentially precarious budget position subject to whims.
Support unit (IT/IS)	Brings GIS within existing support structure, budget protection in early (expensive) stages.	Distance from operational needs, difficulties integrating into front-line units.

Source: Adapted from Somers, 1994

Every organization is different, but somewhere there should be a chart or table that shows the various units of the organization and who reports to whom. The table of organization is a key document in any business or government, and at one level this is a decision of where to put the elements of GIS into this table, what box to draw and where to link that box to the management structure. There can be no cookie-cutter answers to these questions because of how different organizations are. And the placement of the GIS after initial implementation is not a final decision; there are many examples of GIS implementation that started out in front-line units and moved into support units as more departments got involved. Although there is not much discussion of this in the literature, an organization might take a phased approach to locating the GIS by beginning in one or two front-line units with eventual movement into a support or strategic location.

The type of support a GIS needs is little different from other IT implementation. A simple example is if IT training has always been considered an organizationwide responsibility and the budget for that training has always been in, say, the human resources department, that is where money for GIS training will go and that department will be responsible for managing those dollars. But if each unit has always submitted a budget line for training every year or quarter, you will have more complex decisions about who gets how much of the training budget. In the former case all training money would go to human resources, but in the latter case some manager will have to decide who gets how much training. The same would apply to cost items such as hardware and software. There are probably countless examples of organizations that formerly allowed separate units to purchase hardware and software from their own budgets and then brought those functions into the IT/IS department because the variability in software was hampering collaborative operations, and the individual units could not get the same economies of scale on hardware purchases.

Although it is premature to make all the decisions about organizational location in the design phase, you do need to make some basic decisions so that the
people in the organization can see where things will be when the implementation
process is complete. It is difficult to motivate people to build a complex database
and set of applications if they have no idea where it will live after implementation.
Nor are decisions made at this stage irrevocable. You may intend to out-source
application development but hire new staff during the implementation process
with the needed skills. You may intend to store data as separate layers, but by the
time implementation is completed, the selected software vendor has developed a
new and much better data structure, so you adapt to that as well. But deciding
where to put the key components of the GIS in your organization should come
early. Table 9.3 lists some functional groups and generalized arguments for or
against centralizing these functions.

Then there is the culture of the organization to consider. GIS is rarely the first
information technology system to come through the organization; it has probably
dealt already with the implementation of databases, word processing, spreadsheets,
and other related systems. These systems are not GISs, but the issues of integrating
them with the organization's structure and workflow are very similar. If an organization allowed all the units to select whichever word processing software they
wanted, as often happened in the early days of word processing, this would create
real problems of circulating documents around the organization. And certainly different units have existing information systems that work on different operating
systems, have radically different data models and structures, and generally don't
interact well with each other. How the organization dealt with these systems will
condition how they deal with GIS; it is never a completely new thing. Some functions, like software licensing and hardware purchases and maintenance, have such
economies of scale that arguing against centralization of these functions is difficult.

Table 9.3 Centralized or Decentralized after Implementation

Purchase and maintenance:

Software	Makes sense for one unit to manage all GIS software licenses for the organization.
Hardware	Economies of scale from centralization.

Purchase and maintenance:

Spatial data	Specialized techniques and hardware suggest centralization.
Attribute data	Depending on data quantity and variability, decentralized maintenance is likely.
Application development	Central programmers may not understand users' requirements, but de-centralized developers may lack skills.
Training	No clear choice; training needs will vary widely.

For other activities, such as development of applications, the arguments could go both ways.

As with any new project or set of tasks that an organization undertakes, the decision of how much to concentrate the activity in one unit and how much to spread it around to many units is key. GIS is generally thought of as an integrating technology. Because so many users need the geographic information, the GIS and its design and implementation process represent a real opportunity for units to work together for common goals. This often happens, but there are also cases where a GIS can be effectively concentrated and controlled in one organizational location and provide useful information for the other units. Both models can work and neither is inherently better than the other.

How They Did It — Fitting GIS into the Organization: Town of West Hartford, Connecticut

West Hartford's champion in the GIS process was a civil engineer, Bill Farrell, who, even though approaching retirement, saw the advantages for the community of spreading GIS to multiple units beyond engineering. The engineering department began in the 1980s using the somewhat primitive DIME files (precursor to the Census Tiger files) for mapping pavement conditions and as a tool to improve pavement maintenance scheduling. Engineering and IT departments had to work together because of a tightly managed mainframe and difficult to interpret software and data structure. At this time GIS was only lightly being used by this one department, engineering, with significant support from IT.

Engineering conducted a feasibility study for GIS, still on the mainframe at this time. The entire governmental structure was being reorganized at this time, and information technology functions were being moved from the finance department and being placed in administrative services, a move that was designed to make computing capability available to all departments and not just centered in finance. The town also decided, not on the basis of the feasibility study, to go to client/server architecture and allow the using departments to be more independent with the computing needs.

Although these two streams were converging in the town, the state government was beginning a pilot GIS study with the towns in a water and sewer district (the Metropolitan District Commission, or MDC) to examine the effectiveness of a municipal GIS based on the MDCs data and expertise with GIS. With this study under way and the town's movement away from mainframe computing, the engineering department shelved the feasibility study and became active in the state-sponsored GIS project through the MDC.

This project provided software, hardware, digital orthophotography, planimetric data, and training; West Hartford was one of several communities in the MDCs service area that participated. The project was implemented exclusively in the engineering department, but word of GIS and what it could do was leaking out, on purpose, from that department, largely through the work of interns who did GIS work for different departments using the engineering department's equipment, software, and expertise base. During this time Bill Farrell was actively touting the value of GIS in interdepartmental meetings and generally building an attitude of "you've got to have this."

Expansion of GIS into several other departments, principally through the intern relationships, helped the town managers realize that they could provide better services, eliminate internal dependencies for data, and use Internet and intranet technology to do all this. They focused on GIS as a distributive enterprise technology to perform these functions and moved GIS from engineering into IT. Reorganization and a golden hand shake retirement left a position open in IT that required not only GIS skills but general PC, database, and Internet skills, which they filled with one of the interns working in GIS with the engineering department. Figure 9.1 is summary of how this happened in West Hartford. The timing was critical in this example because the possibility of a staff position occurred at the same time management was revamping IT anyway, there were state and regional GIS initiatives that saved a significant amount of start-up costs, and the groundwork had been laid within engineering and around the town. It almost seemed natural when the intern moved into her position. She has since move to another job, but the position was filled, and GIS is firmly entrenched in the working of community government in West Hartford.

There probably is no such thing as a standard process whereby GIS moves from a single department where it got started in the organization to a centralized location where an enterprise GIS is best served from. Figure 9.1 shows how this community went through the process from departmental initialization to placement in the IT department. In the case of this community, it would have been unlikely that GIS would have been initially placed in IT. The experience of the engineering department over a 10-year period was central in the eventual movement of GIS into IT, and that department was very supportive of that move, even though it meant, effectively, the loss of a staff position.

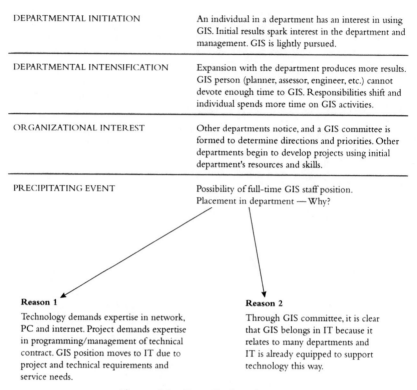

DEPARTMENTAL INITIATION	An individual in a department has an interest in using GIS. Initial results spark interest in the department and management. GIS is lightly pursued.
DEPARTMENTAL INTENSIFICATION	Expansion with the department produces more results. GIS person (planner, assessor, engineer, etc.) cannot devote enough time to GIS. Responsibilities shift and individual spends more time on GIS activities.
ORGANIZATIONAL INTEREST	Other departments notice, and a GIS committee is formed to determine directions and priorities. Other departments begin to develop projects using initial department's resources and skills.
PRECIPITATING EVENT	Possibility of full-time GIS staff position. Placement in department — Why?

Reason 1

Technology demands expertise in network, PC and internet. Project demands expertise in programming/management of technical contract. GIS position moves to IT due to project and technical requirements and service needs.

Reason 2

Through GIS committee, it is clear that GIS belongs in IT because it relates to many departments and IT is already equipped to support technology this way.

Figure 9.1   From Engineering to IT.

Designing the Data Flow

Complex organizations that need to make spatial decisions such as where, among all the possible locations, to put this particular thing or how to assign these resources/people to these places. These decisions often involve many people and units of the organization in that decision-making process. Whether it is the flow of paper and permits in a local government that produces a decision of whether to approve or reject a subdivision proposal or the process used by a fast-food business to open or close a new set of restaurants, there will be lots of people and different units involved. The key decision you need to make in the design and planning phase is whether the GIS is going to be used to support the existing process or is it going to be completely integrated with the process and effectively replace the existing process. If you take the first approach, you will need to design and implement the right applications to produce the reports and maps that are needed. Taking the second approach of directly linking the GIS with the process so that the decision makers can share the information across networks requires some thought in the design stage.

The first approach, using the GIS to produce the necessary output, is easier and provides an opportunity to think about improving the quality of information used in making the particular decision. A GIS usually can deliver different sets of products than you commonly use. For example, it becomes easier to produce mapped output at different scales, with information depending on the scale. In deciding where to locate something, a regional-scale map showing relevant situational features (e.g., competitors, sources of demand) might be needed, but you could also produce a site-scale map with a digital orthophoto behind it to get a quick feel for any potential site problems. The spatial query capability of a GIS makes it much quicker to produce reports based on nearness. GIS has had a tremendous impact in retail site location. The old analog methods of determining sales potential and analyzing potential locations were so time consuming that it seriously restricted the number of sites that could be evaluated. With a GIS this analysis, although still not trivial or simple, can be done much more quickly, which allows more time to consider information that may be harder to quantify but is very important. You would want to make sure that the mapped and tabular output from your GIS was at the right scale and in the correct format to support these decisions. The GIS becomes the engine that produces reports previously done by hand. The GIS may allow you to produce and report information in ways that you could not do before, and this information needs to be produced at the right time in the decision process.

In the second approach, where your GIS is completely integrated with the decision process, you need to redesign the entire process. You need to develop procedures for getting data out of existing spreadsheets or databases or for replacing the components of those databases with new tables in your GIS. Although it may seem that this level of design detail is coming too early in the process, remember that designing a GIS is more like designing a house. You need to set out the specifications very clearly to be able to get accurate bids on what it will cost to produce it. If you know exactly what fields you will need in what tables, where in existing data tables the current information is, and what new fields you must create in these tables, you have a complete template for construction of the database. The decision processes you plan to integrate with your new GIS determine what information you will need. One of the reasons that implementing a GIS seems so daunting to many organizations is that it forces you to rigorously examine your current procedures and data tables. In that examination you often discover that the existing structure is inefficient and the current data are really not adequate for the decisions you have been making. During the GIS implementation process you get an opportunity to improve your processes and develop more useful data for decision making.

Examining the current flow of information that supports decisions is part of the needs assessment. In the design phase you can take time to reflect on that flow of data and see if it can be improved on with the GIS. Implementation of a new information system like a GIS is not just adding another layer to an already complex situation. Instead you should regard it as an opportunity to improve and simplify the decision-making process in your organization. How you deal with the organizational design issues is more important in attaining those goals than is database design and software or hardware selection. GIS implementations usually do not fail for technical reasons but because the organization was not prepared for the technology and the implementation planning was poorly done or not done at all. The new information system must fit within the organization, and the organization must change sufficiently to make use of the new information. Otherwise you will have failure, and failed projects are difficult to resurrect.

ADDITIONAL READING

Curley, K. F., and L. L. Gremillion, (1983). The role of the champion in DSSimplementation. *Information and Management*, 6:203–209.

Forrest, E., G. E. Montgomery, and G. M. Juhl. 1990. *Intelligent Infrastructure Workbook: A Management-Level Primer on GIS.* A-E-C Automation Newsletter: Fountain Hills, AZ.

Huxhold, W. E., and A.G. Levinsohn. 1995. *Managing Geographic Information System Projects.* Oxford University Press: New York.

Kraemer, K. and J. King. 1989. *Managing Information Systems: Change and Control in Organizational Computing.* Jossey-Bass: San Francisco.

Somers, R. 1994. GIS organization and staffing. Urban and Regional Information Systems Association Annual Meeting Conference Proceedings, 41–52. odyssey.ursus.maine.edu/gisweb/spatdb/urisa/ur94004.html.

Somers. R. 1995. Learning from other organizations' GIS strategies. *Geo Info Systems* 5(7):15–16.

Somers, R. 1997. GIS management strategies and issues. *Proceedings of the Spatial*

Information Research Centre's 8th Colloquium 1–7. University of Otago: New Zealand
divcom.otago.ac.nz/conferences/geocomp97/CD-ROM/sirc96/papers/key1.pdf.

Somers, R. 1998. Developing GIS management strategies for an organization. Journal of Housing Research 9(1):157–179.
fanniemaefoundation.org/programs/jhr/v9i1-somers.shtml.

Early Management Concerns: Interacting with the System

The database is complete, the hardware and software installed, and the applications developed and tested; your organization's GIS is ready for use. However, you are really only part way through the first round of a cyclical process, the database development cycle (see Figure 10.1).

It may appear that the process is linear, but it is actually circular; after implementation you need to think about evaluation and planning for system maintenance. But before these steps, you need some experience working with this new system you have created, which means that staff must be trained.

Before you can start training staff to use the new system, you must outline the various roles that people will adopt when they use the system. A role is a defined set of database operations and access. Looking at roles rather at individuals is a better approach for several reasons. First, people come and go in an organization, but a role is relatively constant. Secondly, individuals will often have multiple roles in an organization; for some activities a staff member may be a front-line employee doing specific work on a project, but on another project that same individual may be the project manager with the additional responsibilities attached to that role. Nor is a role exactly the same as a job description, which may also contain multiple roles. Defining the roles allows you to control access to the database, provides a structure to ensure security of the data, and guides you in setting up the training that the staff will need to make good use of the new system. Although there may be many subroles within these general roles, the set of roles consists of viewers, spatial data modifiers, attribute data modifiers, application developers, and administrators , as shown in Table 10.1.

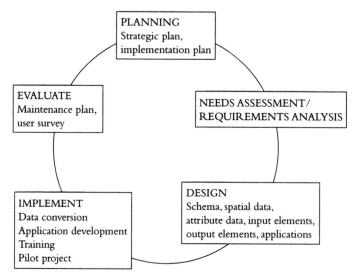

Figure 10.1   The Development Cycle.

Table 10.1 Roles	
Role Classes	*How Does This Role Interact with the GIS Database?*
Viewer	Sees only that portion of the database required for the task(s). Has no privileges to change information or the application(s) through which data are accessed. Needs some understanding of the database structure; mostly needs training to run specific applications.
Spatial data modifier	Able to insert, delete, and modify spatial features (points, lines, and polygons) in the database. Must understand the structure of the database and the relationships that exist among features and layers.
Attribute data modifier	Works only with views into the attribute data necessary to modify the attribute features controlled. May also have viewer role for the spatial data. Spatial data modifier may have this role as well.
Application developer	Will be a viewer as well but may not need either modifier role. Must work with other roles to develop custom applications to speed and improve workflow.
GIS database administrator	Must understand the GIS database structure but need not have any of the other roles, but viewer would be desirable. Functions will not be much different for GIS database compared to other databases administered.

User Roles

Viewer

The simplest role is that of viewer. A viewer needs to understand only the interface to the system with which he or she will be working. In many systems this will be a simplified interface linked to a particular application or set of applications and not the full set of buttons, tools, and menu items that would normally appear in the software's interface. Most GIS software systems allow you to create simpler interfaces by removing functionality from the standard interface or by constructing sets of interaction tools from scratch and building custom interfaces using programming tools supported by the software. This is how the original software vendor builds the full interface. Object-oriented programs are built in pieces with snippets (or large pieces) of programming code associated with buttons, tools, and menu options. The core functionality is called from already constructed bits of programming. The first approach to customize the interface to support a particular user application, by using a full license of the software and removing functionality from a viewer, may appear expensive and backward at first. Why would you want to pay for a complete license of complicated software when you are going to take away most of what it does to simplify the interface for a routine application? The answer to that question is that it may be quicker and less expensive to do that than to develop the custom application from scratch. A GIS installation with a relatively small number of users may be better off using full software licenses even for simple applications and removing the unnecessary functionality to create a simplified, customized interface for the viewers. On the other hand, a system with many users would prefer custom interface development. This approach creates custom interfaces from scratch with only the required buttons, tools, and menu options for the application. This is especially appropriate when many viewers will use the same set of applications. Whether you simplify full-license interfaces or create custom interfaces from scratch is unimportant to the viewer, however. The viewer just needs to know how to do the work, which buttons to push, which menus to pull down, and which decisions to make in dialog boxes that force choices. Much application development consists of taking standard functionality and rearranging it so that specific tasks are easier to perform.

Usually you want viewers to have little or no control over the interface or the data they are viewing. Locking out a viewer from modifying data is critical, especially if everyone is working with a centralized database; if a viewer is working with a copy or partial copy, it is not as serious. You might even want to lock a viewer from altering how data are symbolized, but there may be reasons for allowing a viewer to change the colors or symbols used to identify different features. Some users like to be able to customize the application a little bit so they feel some ownership, so you might want to allow them to modify some design elements. Some viewers may need, or want, more control than others. But the pure viewer role does not include any ability to modify data. Viewers are users who routinely go into the database with standard queries and produce standardized output in a specified format. The abilities to modify data belong to the next roles.

Spatial Data Modifier

Modifying spatial data can be a relatively simple process. An application that takes the address of a client, matches the address with a pair of geographic coordinates, and adds that representation of the client to the database is not complex due to the simple geographic nature of a point. A point needs only a unique identification number, a pair of coordinates, and possibly an elevation to completely describe its location. For example, the entering of the location of a police call is a spatial modification process. A user with the spatial modifier role must have access to the data layer of police calls and appropriate permissions to add features and unique identifiers to these new point features. Additionally, the application should have some process of quality control even it is only as simply as on-screen verification of the content, checking of the database to see that the unique identification number is truly unique, and providing an opportunity to shift the location of the feature before committing it to the database.

If the features are best represented by lines in the database, the level of complexity in spatial modification goes up, and the role gets more involved. Lines usually need to be linked up with other lines to be useful features. This means that the spatial modifier must check to see that the new line being entered is linked directly to existing lines if this is a requirement of the application. Unless the linkages between linear features are explicit, routing applications will not work because there is no way to move from one line to another in the system. And just because it may appear on the computer screen that one line is linked to another, it may only touch the second line and not actually be snapped to it (i.e., there is no intersection feature or node defined). Composite linear features such as highway routes and pavement management sections that are made up of several separate linear features take even more care in their insertion into geographic databases. An example of an application for a spatial modifier working with linear data is the person responsible for maintaining the coordinates of a street centerline file for an area. The line needs to be inserted and its location checked, and it needs to be snapped to other lines in the network.

Polygons or areal features are composed of lines and possibly more complex elements such as arcs (sections of a circle) or splined linear features, which makes their construction more involved. If the database supports topology, anyone modifying spatial data must take care and know how to rebuild the topology after modifying the database. Polygon data are the most difficult to maintain and update in a GIS, particularly if your system requires and supports the topological vector data model. This means that as polygons are added, deleted, or modified, their adjacency to other polygons must be maintained or created. Also, the spatial modifier may have to add label point features where labels from attribute information will be drawn and possibly annotation or text information that relates to the new or modified feature. An example would be the user whose responsibility it is to process splits and combines of land parcels in a land parcel layer, a common and important task in any governmental unit that is responsible for land assessment and taxation.

Attribute information about geographic features is easy to explain to users; the structure of a table where the rows represent the features (utility poles) and the columns stand for what we know or want to know about the features (date of installation, ownership, etc.) is simple. But the structure of spatial data is more complex, and modifiers of spatial information need to understand exactly how the system stores and retrieves the geographic features. Although viewers need only to understand the interface and how it works, spatial modifiers need to understand the data structure and how the system stores information.

Deleting or moving spatial features in a GIS database is relatively simple. To delete a feature you merely select it with some tool (the screen displays the feature differently because it is selected), and then you hit the Delete key. The system may prompt you to state that you really do want to remove the feature by clicking Yes to a question, but then it will delete the feature. And, unlike many RDBMSs, that feature is completely deleted, not just marked for deletion and removed the next time the database is compressed or backed up. Working with spatial data can have real consequences, and spatial data modifiers learn early to work on copies of the database or portions of it to make changes and then to insert those changes into the master database after modification. To move a feature, you usually just pick it up with tool and move it to a different location and drop it there. The attribute information stays the same; all that changes is the location of the feature.

Inserting new geographic features in the database often requires special tools such as GPS receivers, scanners, and digitizing tablets. More and more information is coming into GIS databases through GPS, and there are lots of issues of accuracy and processing that you need to know before you can do this. How accurate will the locations be? To record a point feature such as a road sign, will we take multiple points and compute the average, or will we just take a single satellite fix as the estimated location? After taking the points in the field, will we use differential postprocessing to improve accuracy? The technical questions around using GPS to bring even relatively simple point information into a GIS are considerable. Scanners have similar technical questions of resolution, accuracy, and the complex task of taking a scanned, large set of pixels that are defined in the local coordinate system of the scanner and transforming those grid cells into a much smaller set of point, line, or area vectors in a geographic or projected coordinate system that will allow the new data to fit with your existing database. Just about every novice GIS modifier of spatial data has taken information in some kind of local coordinate system such as the scanner or digitizing tablet and tried to overlay it with data in an existing projection only to find that the new data sit way down in the lower left-hand corner of the map view, near the 0,0 point of the coordinate system, whereas the projected data it should overlap with sit in the upper right right-hand corner. Digitizing paper maps, whether on the screen with heads-heads-up digitizing or from a digitizing tablet, present similar problems that take quite a bit of technical understanding and patience to solve.

Users and managers who do not understand the structure of GIS databases often gloss over these questions by just assuming that GPS or scanning paper maps will somehow magically result in useful, accurate data. Those in the role of spatial modifier are the ones who will actually do the necessary transformation, and they need to understand how to do it and ensure that new data are correctly inserted into the database. It is remarkably easy to do it poorly and quite difficult and often time consuming to do it well. It is always more efficient to take the time to do it correctly the first time than to try to fix it up later, however. So unless the modification of spatial data is simple, as in the police call example earlier, this role requires considerable training and, often, specialized equipment, so usually a relatively small number of people in the organization will have this role. One final concern for the spatial data modifiers is with the spatial indexes. Most GISs will automatically update the spatial index tables when features are removed or added or their location is modified, but some require explicit rebuilding of the index tables.

Attribute Data Modifier

Attribute data modifiers may appear to have little or no direct interaction with the GIS database because they can often perform their tasks without any map-like interfaces. But their role is critical to the maintenance of a useful set of data. Because they do not interact directly with the spatial information and may not even have the role of data viewer, they may not know (or care) that they are dealing with a GIS. In a data set where spatial change is infrequent relative to attribute change, there may be many people who have this role. A property assessor's database in a highly developed large city with few new lots being created is an example of this combination of change frequencies. Compared to splits and combines of lots (spatial data), changes of ownership and assessment information (attribute data) are much more frequent. Those changes may be processed in a database management system using sets of input screens that contain no spatial information at all; access is by address or property identification number. Users with little or no GIS experience can do the management and quality control of these changes to the database. There is one important caveat, though. The attribute modifier role must not be allowed to add or delete features in the attribute data set without working with a spatial modifier. In GIS databases that are working with layers rather than fully integrated databases, the GIS software manages the link between the spatial and attribute data, but the data reside in different physical tables. There is one very widely diffused GIS that, for very good reasons, will allow the deletion of an attribute record in a data table without deleting the corresponding data record in the spatial table. So when your view into the data is through the attribute data table and you delete a record or feature from that table, the corresponding record is not removed from the spatial data table. But when your view into the data is as a spatial modifier and you remove the feature from a map-like view into the data, the corresponding record in the attribute table is removed. The reason this is allowed is that you often have data tables in your applications that have no associated location information associated with them. Modification only of the attribute information without deleting or adding geographic features requires no interaction between the spatial and attribute modifiers. To summarize, if attribute modifiers are working with spatial data, even if they have no need to see the mapped view of the data, they should have the permission to delete records in that table,

Application Developer

Application developer is a role that does not require the ability to modify the spatial or attribute data, unless that is the purpose of the application. An application developer works on procedures and software code that facilitate the other roles. Increasingly, core GIS software systems are designed so that developers may write applications in high-level programming languages that access procedures and libraries inside the GIS core software code. Some systems advertise that their software can be manipulated and extended using any software built on Microsoft's Common Object Model (COM). These so-called COM-compliant software systems allow a developer to write application code in a standard programming language such as Visual Basic for Applications, (VBA), Visual Basic (VB), C++, and Java. This application code makes use of extended functionality for these languages that deals with spatial data. Earlier GIS core software sometimes included a proprietary scripting or programming language for applications. This required an application developer to learn this new language to customize the GIS. Now application developers already skilled in a particular language need only to understand how to use the extended functions and subroutines and the data structure to program the GIS to do additional tasks. This is a real advantage for organizations that already have programming staff that have been working on nongeographic applications involving other COM-compliant systems such as word processors, databases, and spreadsheets. The programming interface will be exactly the same, as will the process of developing code for the application by working with the person or group that needs the application and writing, testing, deploying, and updating the code. The industry is quickly moving away from proprietary languages for applications and toward using these standard programming languages. This means that programming your GIS to do new tasks becomes almost exactly the same as programming the other software your organization has been using. Your original implementation plan included certain applications that were to be automated in the GIS, but there will be a continuing need for new applications and improvements of existing applications. The bulk of the cost of the initial construction of the system is in the data development and conversion, but it is the continuing development of applications that makes the system more useful.

Application developers have the most complex role in the system. Not only do they need to understand the software and how it works and the database and how it is set up, but also they need to be able to interact with the users who need the application so that they can create a useful product. Although many programmers get a little tired of sitting in a cubicle and cranking out code, writing applications for a complex GIS installation can be very interesting work. The implementation plan should outline the order in which applications should be developed, and sometimes it is difficult to keep to that list of priorities. Users will be quick to request modifications of existing applications and bring pressure for new ones to be written. So it is important to have a procedure for allocating the time of application developers. If the database is well designed and maintained, most of the growth and development over time will be in the applications. For each new application or change in an existing one, the product development cycle begins anew; it just does not take as long to work through the process. It must begin with planning and move to assessing the needs, designing the application — including the process, data required, and output — implementing it, testing it, and having the users evaluate the final product. Although the development cycle for the entire GIS may take years to work through, it may take only a few weeks for new applications to go through. The cycle gives a structure to the process, and experienced programmers are used to this cycle.

GIS Database Administrator

Database administrators (DBAs) are responsible for the overall database, and in this respect a GIS database is not much different from any other database that the organization may be using. A DBA may set up roles in the system, back up the database, monitor its use, see to its overall security, and generally be the guardian and administrator of the database. To good DBAs, a GIS database is simply another database that puts requirements on them and will not appear to be anything special. If an organization already has multiple databases to support other activities of the organization, the GIS database becomes just another administrative task similar to others. DBAs do need to be familiar with the data model, data structure. and file structure of the core GIS software and how it is linked with other data in the system, however.

If the organization is small, the current DBA may also serve as the GIS DBA, but in larger organizations there is often a need for a separate individual whose responsibility, part-time or full-time, is administering the GIS database. In this situation, it is useful to have a division of labor between the two roles of overall DBA and GIS DBA so that there is no duplication. For example, the enterprise DBA will have procedures for setting up new users in the organization's technology system that may involve training and testing. If the new employee is going to work with the GIS database, it may be the responsibility of the GIS DBA to set up the user role(s) and do the training and evaluation on the GIS. Perhaps the system DBA would be responsible for setting up, tuning, and evaluating indexing systems on attribute information, but the GIS DBA would have the same role with respect to the creation and maintenance of spatial indexes. All organizations are different, and larger ones tend to have a larger number and more specified roles within the GIS, so separation of responsibilities is important.

Managing User Roles

As your system matures, the number of possible roles for interacting with the system will grow; the set of roles is not static. Within the viewer role there will be subroles that may allow viewing and browsing over larger or smaller portions of the database. The spatial modifier role may have many subroles depending on the distribution of responsibility within the organization for various layers and sets of features (e.g., a planning department may be assigned the role of spatially modifying the zoning layer but not have rights to touch the layer of land parcels). You may even choose to assign some maintenance roles to outside organizations that will perform the maintenance and updating on a contract basis. Many consulting firms will willingly take on many of the more complex and involved roles, leaving your staff only to work with the applications. Often the consulting firm that was most heavily involved in the database design and implementation phases will be willing to engage in these kinds of maintenance contracts. It is not uncommon, for instance, for the firm to leave some of its staff in the implementing organization at least part time during the early management period of the GIS.

Earlier in the book we discussed the relative merits of GIS databases constructed of data layers and those that adopt a SERD model. If you have adopted a SERD as your GIS data model, security and role creation are integral functions in the database management programs. But if you construct a layered database, it will be more difficult to keep the database secure and control access to it. In a database with a layer structure, data reside in many different folders, and your principal means of access control is through the operating system, which is more difficult than controlling through a database management system. In a SERD all the data are inside a single database, and there are management tools to control access to the tables in the database. However, it is not uncommon for an organization to implement a hybrid GIS where some units interact with a layered database, others with a SERD, and there even may be more than one SERD in use in the organization. One organization had existing databases that used three different database management systems and wanted to link its existing data with the new GIS data. Obviously, there is no clear, cookbook-like process for managing access to the data, but the costs of failing to do it well are considerable.

Managing the roles and the associated access controls is important for all GIS databases but especially important for databases that undergo frequent modifications from multiple users. A local government that processes a relatively small number of land transactions over a month may be able to manage the process of changing the database very simply by making occasional changes all at once. A large city or public utility, however, may have many concurrent users whose sole task is to modify attribute and/or spatial data, and controlling the quality of that kind of interaction is very important. A common strategy is to make all changes to temporary files and have a manager or supervisor review the changes before committing them to the database. A second strategy involves the concept of versioning. In this approach each user works on a version of the database and makes changes to that version. The software checks those changes to see if they conflict with changes made by another user. In a large organization it is not uncommon for two people to be working on the same data simultaneously, requiring a process to reconcile any differences between their two versions. When versioning is being used, changes are not committed to the database until someone reconciles the versions or tells the system which version to insert.

It is clear from this discussion that the really important task in managing roles is how you establish procedures for modifying the database. These procedures need to be documented and understood by all modifiers of spatial or attribute data. Because the organization has been working well without a GIS and has existing databases and procedures for modifying them, the only task is to insert the appropriate quality control steps for modification of any new data that will be managed in the GIS.

Managing Desktop Interfaces

Research in the area of how people interact and actually use a GIS has been going on for almost a decade, but the published literature is not too useful to day-to-day practitioners. Over time the improvements to how people interact with a GIS will move from the researchers to the applications designers and software manufacturers. Early GISs had what now seem intimidating black screens with a prompt that just waited for the user to type in a command. These interfaces were copied from the Unix operating system where a command line interface with a structure likes *Command [required parameter1, required parameter2...] {optional parameter1, optional parameter 2...}*. These interfaces have now all been replaced with GUIs based on Windows icons menus and pointers (WIMP). The acronym that was chosen for these interfaces says a lot about what the designers thought about the users. There is work going on to improve these interfaces, but today most GISs present both the novice and experienced user with a bewildering array of pull-down menus, buttons, and toolbars. Users are able to modify the appearance of the interface by removing objects, creating custom toolbars, and placing frequently used objects conveniently on the interface, but the standard interface is still an intimidating thing to new users.

Most users will be interacting with the system through a GUI. GIS software often also has command line functionality, but only skilled power users feel comfortable with a blank screen and no buttons to push or menus to pull down. A management concern is how much customization of interfaces to do. The software designers wanted the maximum number of tools and functions available to users, so the menus, button bars, and toolbars are crowded with options. A lot of training time can be consumed explaining things to users who may have no need for that much functionality. A lot of programming time can be consumed by having to create a large number of customized interfaces as well. Some customization is relatively easy in that with most GISs you can define custom toolbars by arranging functions you might need frequently on conveniently organized toolbars that can either be anchored on the interface or allowed to float over the data view. By doing this you can move the small set of tools that an application might require from locations that might be hard to navigate to and place them in a usable location. The metaphor is that of a toolbox, and the system allows you to put the tools you need frequently within easy reach. This kind of customization of a fully licensed version of the software is not difficult and can speed work significantly. Figure 10.2 is an example of this type of interface; most desktop GIS GUIs have a similar structure, although the icons, menu options, and tool arrangement will differ.

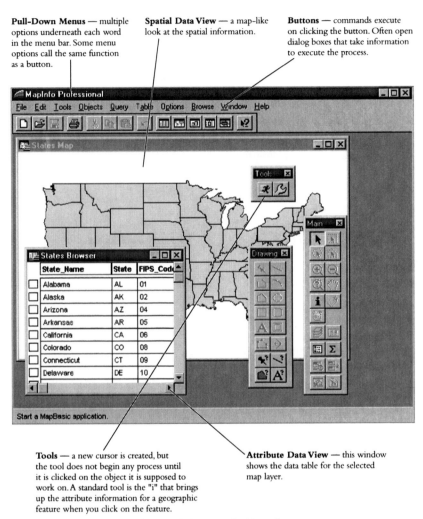

Pull-Down Menus — multiple options underneath each word in the menu bar. Some menu options call the same function as a button.

Spatial Data View — a map-like look at the spatial information.

Buttons — commands execute on clicking the button. Often open dialog boxes that take information to execute the process.

Tools — a new cursor is created, but the tool does not begin any process until it is clicked on the object it is supposed to work on. A standard tool is the "i" that brings up the attribute information for a geographic feature when you click on the feature.

Attribute Data View — this window shows the data table for the selected map layer.

Figure 10.2 A typical interface.

Creating custom screens from scratch is more difficult and requires programming skills in a scripting or programming language the software supports. Using this approach, application developers create interfaces from the ground up rather than rearranging or removing functionality from the entire system. This approach may be necessary for licensing cost reasons. If you have many individuals doing a small set of routine tasks, it is usually less expensive to construct these custom applications and interfaces rather than spending the money for complete licenses. The appropriate license for the tool development software allows you to build standalone programs that have access to the required libraries and subroutines for doing the work. If your implementation has relatively few users, it may be less expensive to equip each user with a full license and simplify the interface for the various applications. Some managers in organizations like to construct custom interfaces because it gives the appearance of being something unique to the organization. Through customization of the screens the core software is hidden behind the interface, and there is a sense of corporate or organizational identity with the system.

Custom interfaces are also excellent tools for helping less technically minded users feel comfortable with the system. All GIS implementations have problems with difficult, stubborn, and sometimes unwilling-to-learn users. Typically these users regard the system as a threat and see it as a statement that they have not been performing well. Otherwise, why would you implement a new system to do their job? These users are the most difficult for managers to deal with in GIS or any technology, and an interface that simplifies the work and makes it accessible to this type of user goes a long way toward solving that problem.

Managing World Wide Web-Based Interfaces

More and more viewer interaction with GIS databases is occurring through Internet and intranet applications. Users gain access to these applications completely through their Internet browsing software (e.g., Netscape or Microsoft Internet Explorer) and require no GIS functionality on the client side of the connection. The software that constructs the maps and processes the requests from the client's browser to the map and data server, however, must have specialized software that facilitates the serving of the GIS database to the internal intranet or external Internet.

In this type of interaction, the interface design is of extreme importance, and decisions about what functionality to place on it and how to design it so that users get the information they need must be made at relatively high levels in the organization because of the face that it presents to the public.

There is a lot of work going on both by the GIS core software providers and third parties around Web-based interfaces to geographic data. Given the time it takes for a book manuscript to actually turn into the object you are holding, anything we say here will probably be out of date when you read it. Serving static graphic images of maps across the Internet was a very early use of the technology and is still an important means of distributing geographic information. Taking a paper map, scanning it into digital form, and then making that digital file available across the Internet is an effective way to distribute already existing maps. This has been especially useful in making historic maps available to many viewers who otherwise would not be able to access these documents easily.

More recently organizations have taken to placing some or all of their GIS database on the World Wide Web for users both inside and outside the organization to query. And people are using these databases all the time. The map-making and direction-finding sites on the Internet such as Mapquest.com and MapBlast are among the most frequently visited sites on the Web. These sites are network-based GISs that, due to their efficient indexing and system architecture, can rapidly determine least-cost paths from any addressable point in the United States to any other addressable point and create a set of driving directions complete with required turns and distances in less time than it took to read this sentence. Yes, people get lost with the directions and sometimes they appear not to be the optimum path to people who use similar paths daily, but they have become an important locational resource.

The design of these interfaces is always much simpler than the professional desktop interfaces that come with commercial GIS and look more like customized, simplified application interfaces. They make use of standard tools such as 🖐 (pan the map to a new location, 🔍 (zoom into a small area of the map view), 🔍 (zoom out to a wider view), and ℹ️ (display attribute information about the feature. The goal is to have the tools and buttons so self-explanatory that users can start getting information from the database almost immediately with no training. Testing and serious user feedback is vital in designing these interfaces. There is a temptation to include too much space on the screen for identifying information about the organization (e.g., large names, logos, disclaimers, etc.), and too little space for actual map and data content.

No matter what programming language is being used for these Web interfaces or what GIS core software or database structure they are built around, they all work in fundamentally the same way:

- When the user accesses the Web page through a browser, a default template page and the server sends a graphic map image to the client browser. This page contains the software coding necessary to communicate with the GIS data server.

- The user clicks on the available options to modify the map and/or access tools to query the map. There is often a list of layers like a legend that can be clicked on and off. Usually the available layers are predetermined for the user, and there is no opportunity to browse for new layers. The software in the local downloaded page takes the information and creates the query in a format the server GIS data serving software will understand and be able to execute.

- The browser then sends that query command to the GIS data serving software, which executes the query and sends back a new map image if the request was for a new map or a file containing attribute information if the request was for data.

Every zoom, pan, or request for modification or data is a separate query and must be sent across the network to the server, which returns either a new map graphic or data. There are some applications now that actually download spatial and attribute data and make the query on the local client machine.

These applications are becoming more and more common, particularly with organizations that have legal obligations to provide certain information to the public. Organizations that implement these Web applications for routine queries notice significant drop-offs in the number of clients who come directly to the organization for the same information over a counter. The service is often provided free, but sometimes the organization will charge monthly or yearly access fees and protect access to the Web site in that way. If you are considering Web access for your data, it is important that you consider the hardware implications, particularly the server and networking issues for intranets, in the early design phase. The interface design can wait until the database is up and going and typically will go through its own development cycle.

GIS Interaction and the Organization

User roles, interfaces, control, and access must be handled at a detailed level, but looking at interaction with the GIS from the point of view of the organization as a whole, you can expect that the interaction will differ as a function of where in the organization the user is (see Table 10.2).

Front-line users have the most direct interaction; as you move through management levels, interaction becomes less direct, and people begin to interact with the output from the system rather than the system itself. There are certainly exceptions to this; at least two U.S. governors have an operating GIS on their desktops and use them regularly to assist in the administration of their states. But usually top management is concerned with the output of ordered information to help them make decisions, and how they get that information is not a major concern. Midlevel management is a bit more directly connected with the GIS, but the connection is tighter with management issues than with the data and applications themselves. Midlevel managers in GIS-using departments have interaction concerns that center on how they can get applications that will help their people do

Table 10.2 Typical Interactions with the GIS	
Front-Line Users	
Public	Occasional, information-driven, simple interfaces to answer simple questions, Web-based applications
Viewers	Regular, application-driven, customized interfaces; thin clients (i.e., programmed secure applications)
Modifiers	Frequent or as-needed, data-driven, complex interfaces; thick clients (i.e., full GIS licenses)
Midlevel Management	
Direct GIS supervision	Occasional, thick clients, process, and report-driven, problem-solving interface and access management
GIS-using units	Application and interface development, work-flow improvements
Top Management	Output; reports and maps, goal-driven, may not directly interact at all

work and how they can manipulate the organizational system to get the GIS to support their work. They may have a license for the software and even be skilled users, but they are not daily producers of work from the GIS. Midlevel managers in GIS units themselves have direct responsibility for the database and the supported applications, so their level of interaction is higher and covers lots of different types.

That daily interaction is between the front-line users and the GIS. This often occurs through thin clients — small bits of programming or Web tools that perform a small set of functions quickly. Front-line users who are data modifiers will probably require thick clients, full software versions, because of their need for more functions and the ability to perform multiple tasks. Thin clients are for simple tasks, and thick clients are for users who often need to be able to figure out different ways to do things.

How different users and units of the organization interact with the GIS will change with time. Users will discover new things they want the system to do, and units that were not considered prime users of the system during design and implementation will find that they want to be able to use it as well. GIS systems have a tendency to work themselves into lots of areas in an organization where the designers originally did not think they would go. So the management of the interaction with the system is an ongoing task. By the time most people in the organization have some level of interaction with the GIS, the system begins to fade into the background. It may have started out as a GIS in one or two units and was their GIS. Then as units came together around the common database, it became the corporate or enterprise GIS. If the growth of the interaction is managed well, eventually it becomes just the GIS or even the system, and it sits quietly and effectively behind the activities it was designed to support. Most people, except those who manage it, don't think much about the information system that generates their paychecks. At least until it fails and they miss one. Eventually a GIS gets to that level of transparency; it is just something that is there, reliable and helpful.

A New Committee

During the design and implementation phases there was a working committee in the organization that oversaw those elements of the process. Now that the system is installed and operating, a new group must be established to deal with the ongoing development and maintenance issues. The first committee, if they did their work correctly, has worked themselves out of any tasks. The GIS management committee has a different set of tasks. Specifically, it does the following:

♦ Sees that the training and applications development specified in the implementation plan actually occurs. Users will be wary of the new system, and the committee will have to work with trainers and managers to make sure that the users understand how to use it and what it can and cannot do for them.

♦ Functions as arbitrators and decision makers when conflicting claims over data or access arise. Even the best design and implementation processes miss important things, questions nobody thought to ask before. This committee is the group that will answer those questions.

♦ Sees that the overall maintenance plan is followed, particularly with respect to system evaluation.

The membership of this committee should have some overlap with the first committee, but that overlap should not be total. You will need a mix of direct users and management from the key GIS-using units in the organization on this committee. The design and implementation process should identify the correct people for this committee. It is important that there be as little a gap as possible between the dissolving of the committee that oversaw the first parts of the process and the formation of the GIS management committee; too long a period of time and there is no oversight. This is a new information system that affects many units of the organization, and there needs to be a formal structure, this committee, to address concerns.

A key responsibility of this GIS management committee is developing and overseeing the maintenance plan for the system. Just like automobiles and trucks, there is regular maintenance that an information system needs to function well, and there needs to be a plan for that process.

Evaluation

At some point, preferably after users have become familiar with the new system, and the kinks, bugs, and so on, have mostly been worked out, it is time to conduct a formal evaluation of the design and implementation process. This is a step often overlooked, and if there is a general level of satisfaction with how the GIS has worked out, the organization probably will not fall apart if this step is omitted. Sometimes top management, particularly those who might not have been the strongest supporters, might insist on it. Here is where extravagant claims about the benefits of GIS for the organization can come back to haunt you. A GIS manager once was overheard discussing the needs for updates on the aerial photography and planimetric information for the GIS with the chief budget officer of the organization. As they were talking about how to fund it, the budget chief wanted to know how much they had saved from the 17 people they got rid of because the GIS was now doing their work. The manager had to quickly go back to their strategic plan for GIS and show the budget chief that no such claim had ever been made. But whatever claims were made in the planning process about how this technology was going to revolutionize the way the company does business and provide almost immeasurable benefits to the company, at some time someone is going to want to evaluate the reality of those claims.

So the GIS management committee is going to have to conduct an evaluation; it is a fact of life. Programs and activities that consume significant resources in an organization regularly get looked at to see if they are still returning value to the organization. Nobody really likes to do it, but everyone accepts the necessity. There are several basic components to the evaluation of an enterprise GIS:

♦ Go back to the implementation plan and document whether or not each of the implementation tasks was accomplished on time and on budget and if not, why that was the case. Showing that the actual implementation of the GIS followed the plan closely shows that your preimplementation work was well thought out and executed.

♦ Survey the users and managers of the departments using the enterprise GIS on their before and after impression of how work is being done. The GIS manager referred to previously had users estimate how many times they did certain tasks and how long, on average, it took them to do the task. He did this before implementation of the GIS and then afterward during evaluation. So he was able to document time and frustration savings brought about by the GIS. All the other techniques of survey research, such as interviews, focus groups, and so on, are useful here as well. If consultants were involved in any stage of the design or implementation process, be sure to include questions about the level of satisfaction with their work. They love to hear that kind of feedback (or should), and if it is positive, it validates the committee's selection of those particular consultants.

♦ Try to come up with some kind of summary time or money estimate of how the new system is paying off.

♦ Include copies of the routine and nonroutine output from the new system. It is possible that some individuals in upper management have not yet seen all the products that the system produces, especially the routine front-line applications.

♦ Make sure you do not whitewash the system and be honest about what shortcomings there are and how you plan to remedy them. The product development cycle is never-ending, and the results of the evaluation get fed right back into the planning for the next stage of development.

There are many other management issues around GIS and lots of printed and Web-based resources to help GIS managers do their jobs better. But like any management activity, sometimes the best sources of information on how to make things work better is through personal contacts with people who are doing similar things. The GIS community, at least in North America, is an amazingly open, helpful, and friendly community. Virtually every state and many metropolitan regions have some kind of formal or semiformal GIS organization that cuts across the barriers of private/public/quasi-public GIS installations, different software users, large installation, small installations, educational organizations, consultants, and software vendors. The software support sites are excellent places to get answers to specific questions about how to do certain things with your software and hardware. These organizations, or more accurately, the people in these organizations, are excellent places to get answers to more difficult management questions like How do I know I need to hire more technicians and how do I make a case for it?

or My staff is pressuring me for more staff development and training to improve their GIS skills, but my production demands are so great that I can't really meet those needs even though I have the financial resources. What do I do? Or a very common management question: I'm about to lose a key staff person to a better job in a week. You just did a search; was there anyone in the pool you didn't hire that might be able to do this job? It is likely that someone in that group has dealt with similar issues and can help.

So the management concerns and the product development cycle continue to move forward. The next version of your chosen software is coming out in a few months, and you have to decide if or when you are going to upgrade to that version. The new version is abandoning the 8-year-old proprietary scripting language and going to use of COM-compliant programming language, and there will not be readily available interpreters for the applications. Your server crashes. The network you thought was large enough to handle the data transfer volume you expected doesn't work quite as well as advertised. On the positive side, management loves the new map products you can produce and the speed with which you can produce them. Every week brings a couple of new people down to the GIS operation (they to ask for some specialized product you hadn't planned on but find you can do without too much work). Three years have passed quickly, and it is time to start planning the next round of aerial photography, and there have been such advances in satellite technology that you need to deal with an entirely different set of questions this time around. These kinds of questions and concerns never stop. Nor do they in any management activity that is centered on the rapidly changing world of information technology.

A well-designed system is a flexible system. If you took the right amount of time and spent the right amount of resources, more than you really wanted to of both, at the time they should have been spent, you should have a system that can adapt and still meet the organization's need while adapting. Constant, planned change is the norm in GIS. The cycles never quit, and before you know it, the organization and the people in it regard the GIS you labored so long and hard over as just another piece of the infrastructure. It has completely melted into the background of the support system of the organization and everyone takes it for granted. That is really the goal: becoming so integrated with the day-to-day working of the organization that its use is no more remarkable than using a spreadsheet or word processor. New York City calls theirs a GIS utility. It is back there working, supporting the organization just like the water, sewer, and electrical power utilities support our lives. Most people notice it only when it temporarily fails to work. If you get to that point, you have designed and implemented a successful enterprise GIS.

Access Controls

Organizations differ radically in how they control access to the geographic data and applications they have created in their GISs. Some control access very tightly for reasons of confidentiality, security, and/or market concerns, and others open their entire databases and applications to everyone in the organization and even the outside world. This variation in control extends within the organization as well, with some units being very restrictive and others very open. So questions of access control and data security need to be addressed.

Traditionally, the GIS community has been very open. Partly as a result of that tendency and to make it easier for people to use it, GIS software as it comes off the shelf has little security built in, other than from improper use of the software. One widely used software system allows you to put a password on a geographic data layer, but that password is written into a text file that describes the project and is not encrypted, almost no security at all. Because there is little security functionality in the core GIS software, you have to deal with access, control, and security from within your computer operating system and the RDBMS you are using to manage your enterprise GIS. Fortunately, both of these components of your system come with strong ability to manage access and security. Through the operating system you can assign permissions to folders, directories, and individual files. Through the RDBMS you can control access to data tables and can lock people out from even viewing selected fields in individual tables.

Control through the RDBMS

If your data are stored on a centralized server, you control access through permissions granted to users and classes or groups of users when you set up their profiles. Because a GIS may have dozens of users and people come and go within an organization, setting up access inside defined roles or groups within the RDBMS is a simpler method of control than controlling access on a person-by-person basis. When a new user is entered into the system, the administrator will attach the roles that person may assume. When people leave the organization, removing them as users strips them of all roles and access ability. RDBMSs even have the ability to temporarily lock out users without deleting them from the entire system. In a local government you might want to define a role of Property_Viewer, which would allow users to view, in map and table form, selected fields of a complex property database but not allow any modification or even querying of data other than to locate a particular property through an address or PIN query. A simple interface attached to that role would allow only those queries, but the Property_Viewer could not query the database to find property based on the name of the property owner. That functionality could be reserved to a role of Property_Query. Roles are usually hierarchical. If, within the category of users that will be dealing with property data you have View, Query, Selective_Modifer, and Developer roles, each role will need the access privileges of the roles beneath it in the hierarchy.

The setup and modification of roles within a database are tasks that database administrators are trained to do; it is really no different for geographic data access than for any other kind of data access within a database. Setting up the roles and the access privileges is the responsibility of the GIS database administrator working with the overall database administrator. Managers of the operational units that need the GIS will be helpful in defining the parameters of the roles that will support their units, but the details of role creation are best left to database administrators rather than line managers.

Control through the Operating System

Access control through the definition of roles can be backstopped by access controls at the operating system level. These types of controls were present on the early mainframe versions of GISs, became easier to manage with the advent of Unix operating systems, and are now present in the versions of the various Windows operating systems. The notion of administrator at this level relates to the hardware rather than the database. Roles and permissions work within the enterprise database, whereas operating system controls work within the pieces of hardware, (i.e., the disk drives on the system). Although database administrators control access to the database, machine administrators control access to the hardware on which the database resides. Obviously, these people need to work with each other. Although role assignment is a good way to control data access, permissions assignment is a good way to control application access. If you locate the applications that run your GIS in folders where only certain individual users have permissions to execute programs, only those users may run the applications. By denying them the right to modify (write to) those files or directories, you disable their ability to change the application; that permission belongs to other users. This kind of control allows you to permit access to certain users during the testing of a new application and then open it to a wider set of users after you have worked out the problems.

If your data and applications reside on central servers, the control is also centralized, but when you distribute data to users' individual computers, control becomes more difficult. All GIS software contains functionality to take in a data layer, perform some kind of query on it, and then export that to be a new data layer. If users access a data layer to which they have viewing access on a server and then create a copy of that data in a folder to which they have write access, they can then work on that copy of the layer and do whatever they want to with it. Removing that functionality from all the users could be difficult if there are lots of users in the organization. It is like a copying machine; you can control the initial access to the document, but unless you control access to the copying machine, it is difficult to stop users from copying the data and then distributing them to people who should not have access. The only solution to this problem is completely limiting access to the user's desktop machine or removing the functionality to create copies of data within the core GIS software. Data are going to get out. Copies and partial copies of data sets are going to be all over the place, and there is a real possibility that people will be making decisions and developing output from the copies instead of using the centralized, controlled data. If you distribute data across a network or through some physical medium like a tape or compact disk rather than serve it centrally, this problem will be worse, but it exists even if you control the data on a central server. Very sophisticated techniques are necessary to completely control the ability of a user to copy data to a different location, but it can be done.

Controlling Public Access

For organizations that used public money to develop their GIS, the issue of how to allow and control access by the public to the database and applications is important. This aspect of GIS data access frequently receives a lot of attention outside the organization. Managers and administrators can tightly control who has access to what within the organization and have a responsibility to do that to maintain the integrity and quality of the data. It is not in anyone's interest to allow unqualified people to modify data. Controlling the public's access to the data is more difficult. This discussion is separate from the question of what you can charge for the information and deals only with what you have to make available and how you do it. Any governmental organization has dealt with the concerns before, and the implementation of a GIS is only another step in this process of making data accessible. All government's agencies usually seek legal advice before establishing public access policies, and if they do not, they certainly should.

There are a number of questions the legal department might ask the managers of the line units using the GIS about public access. The first important ones to ask when deciding what data to make available are: Do you have to provide public access to the information and/or applications? Is it a legal requirement? In the case of most public data the answer is Yes, but there are constraints. You can deny access for certain purposes. A good example of this is the refusal of local school boards to provide the addresses of the students in the school system. Lots of reputable marketers and politicians would love to be able to identify those homes with school-age children to market products or services to the parents or, in the case of politicians, contact the voters in the household about their positions on education. Laudable as those purposes are, it would be rare for a school board to provide that information because of the risk it could get into the hands of people with less lofty goals. On the other hand, a state will willingly provide the list of individuals who obtained hunting or fishing licenses in a given year. In Connecticut, for example, that list in a database format on a floppy disk will cost $14.00, and it will arrive in a week after a telephone request.

A simpler request, which is almost impossible for a public agency to deny, is the request to view the data set, and local and state governments are moving rapidly to provide map interfaces into their GIS databases across the World Wide Web. Because this is becoming so common, the question frequently arises early in the design phase, but it properly belongs in the management phase of the process. If the database is well designed, it can be made available in an Internet or intranet application. Web-based access through browser software is the thinnest client you can use to view the data. It can put information in the hands of a lot of people rather quickly. Private corporations, of course, have no legal requirement to make viewing of their data available to the public because the public did not bear the cost of constructing the data sets. The decision of whether and how to allow public access is a business decision except around the information that state and federal regulations require for publicly held companies. This information deals with the company's structure and financial well being and does not include proprietary information in databases the company uses to create profit (i.e., a GIS database).

The management of access to a GIS database is a balancing act. On the one hand, you want to have as much data available to users as possible because the simple ability to view a wide range of spatial information may suggest new ideas.

Pre-GIS processes were stingy with spatial data because of the time and cost of making the maps. When you can put it up quickly on a computer screen, people almost immediately want to see more information and more layers because they come up with new questions to ask. Generally this is a positive thing for an organization, but you can go a little overboard and add so much information to a screen that the clutter makes it very difficult to focus on the key issues. On the other hand, you can restrict access to only that information that was needed to do the job before the GIS, having the GIS reproduce the exact process used before. If you take that approach you lose the ability to consider new processes that might work better, but you gain the simplicity of an interface that relates directly to the task. And users vary widely in their ability to absorb and use spatial information. Some people do not relate well to maps and want to see their information presented only in tabular form, whereas others enjoy and need to see the where as well as the what. Your applications need to be sensitive to the different ways people process information. Most, but not all, GIS practitioners are visual learners and are almost messianic in diffusing their way of looking at the world. Some users will resist this as well as the design of access to the database needs to meet their needs. Balancing between too little and too much information and meeting the needs of the visual users and the nonvisual users creates interesting challenges, and how well you meet those challenges is a key determinant to how successful your GIS will be.

Managing the System – The Maintenance Plan

Before implementation you should have prepared a strategic plan, a management overview of how and where GIS would fit in the organization, and an implementation plan, a detailed process of how you are going to do it. You need to prepare those plans, in that order, before implementation. The third of the three plans, a maintenance plan, is no less important but can wait until after the GIS is functioning. Table 10.3 shows the principal issues you need to treat in maintenance plan for your GIS.

Applications often need updating and maintenance as you get feedback from users about problems with the way it works or does not work. Additionally, if your implementation is phased, there will be new applications to develop. An agreed-upon schedule for development of applications in new units and the updating or maintenance of existing applications is important so that all units in the organizations feel that the GIS will work for them and no unit feels left out. Large GIS implementations typically will have dedicated staff for this task, and smaller ones may outsource it to consultants. As users in GIS-using units get more familiar and skilled with it, they will be able to make concrete suggestions for improving the application and may even be able to do it themselves. Additionally, you will discover applications that you originally did not plan to implement in the GIS but later realize that it makes sense to do so. In the first few years after implementation this aspect of maintenance will be particularly important, and you should budget adequate resources.

Table 10.3 Maintenance Plan

GIS Component	Maintenance Issues	How do you accomplish it?
Applications	Updating existing applications; creating new ones	Feedback from users, in-house or contract application development
People	Upgrading skills of existing users; training new users	In-house or contract training and workshops
Software	Upgrades and technical support, licenses for new users	Maintenance contracts with vendors
Hardware	Replacement cycle, hardware for new users	User feedback, keeping current with improvements in hardware
Data	Periodic large-scale data replacement	Periodic update of entire service area versus rotating updates of selected portions
Evaluation	Serious assessment of GIS utility and acceptance	Structured user feedback

Maintaining the skills of the users of the system is also key. The implementation plan will contain a description and timeline for the necessary training, and the maintenance plan should contain the same for upgrading those skills and training new users. Training budgets in most organizations are sometimes considered luxuries and are often early sacrifices in tough financial times. GIS consultants especially recognize the steepness of the learning curve for GIS and are eager to offer their services in this area. Most organizations with GIS will have at least a few very skilled users, but they may be poor trainers, unable to transfer their skills and knowledge base to others effectively. The major software vendors have programs to certify trainers, and these people can provide a valuable service at a reasonable cost. Local universities with GIS programs are also good places to look for training resources if your organization lacks them. Whether you do all your training and skills development in-house or contract with outsiders, make sure any training you engage does the following:

- Involve your organization's data; avoid training with canned databases that are unlike those your staff will work with. Too much training happens with canned data that has little or no relevance to your organization or applications. The exercises are fun, and trainees learn the buttons and menus to do work, but unless they are good at transferring skills, they will find it difficult to move them over to their work.

- Be geared to your application needs and specifically involve your applications. Specific tasks that they will have to perform are what they should be trained on.

- Be paced appropriately for your staff. The software vendors prefer to run multiday training sessions either at your workplace or at their facility because it minimizes travel and support costs for the trainers. But phased training where the staff can get a section in a day or half a day, go back to the workplace to practice, and then return a week later for another day of training allows the necessary time for the material to sink in.

Most consultants who implement GISs for organizations will tell you that there is great reluctance to spend resources on training. These systems are complex, and it takes a while to learn how to operate them, so some kind of plan to get staff up to speed on the GIS will yield significant dividends.

Software also requires maintenance. Initially, most people think GIS software is like word-processing software; you buy the most recent version out there, and when the update comes along you evaluate whether or not you need it and either do or do not upgrade. Some GIS software is like that as well. But the core software used to construct and maintain the database is typically licensed rather than purchased, and you have to maintain up-to-date licenses to be able to use the software. There is an initial significant purchase, but that commits you to annual maintenance fees, typically one-quarter to one-third the original purchase cost. Information systems or technology department staff are knowledgeable about such contracts and usually find them valuable largely because the contracts come with technical support that the IT or IS department is not able to deliver. Additionally, it qualifies you for any upgrades or patches that come along while your maintenance contract is in force. Although software goes through a complex development and testing process before it is released to the public, it is cost-prohibitive to work absolutely every bug out of the system ahead of time. So just about the time you are beginning to get used to the software version x.1, some important patch becomes available to fix some problems that have come up. They may not be problems that affect your organization, but you will need to budget time and resources to install these upgrades and patches. This is an ongoing task, and although it is easy to ensure there are sufficient financial resources, it is also easy to forget about staff resources that may be required for installation and upgrading.

Hardware maintenance and replacement cycles, along with software maintenance, are issues that organizations have been dealing with since computers became useful in business and government. Generally, GIS software and hardware present no real new concerns. IT/IS people not familiar with GIS are often surprised at the storage requirements of GIS databases. As users start working with data, keeping local copies, and trying out new things, the amount of data that will accumulate is impressive. So it is a good idea to build in the purchase of additional data storage capacity in the maintenance plan. Even though the implementation plan might have recommended a certain amount of space, it has a way of filling up quickly. The need for some users to visualize the database requires a significantly higher investment in graphic processing and larger computer screens. The need for large-format output from plotters means more expensive output machinery, which should also be under maintenance contracts. Most organizations already deal with this issue of maintenance contracts for important pieces of equipment, and GIS output equipment is just another example. Sometimes it makes sense to place some computers under maintenance contracts, but in other cases it is more sensible to plan for a 3- to 4-year replacement cycle for the equipment. Maintenance contracts on the computers that actually serve the data are critical, however.

As we stated early, applications drive the entire process, and the data that support the applications drive everything else. So the timeline and budgeting for maintaining the GIS database may be the most important element in a maintenance plan. Usually, at least 80 percent of the total cost of GIS development is in the data, so it is reasonable to assume that a similar percentage of the maintenance costs will go to the data as well.

Often a lot of the data development and conversion in the implementation process is done by consulting firms that deliver the newly formatted data from legacy databases so that it is more accessible to the GIS. Through normal, routine processes you make updates and modifications to that data. But sometimes you need to reacquire significant pieces of the GIS database because they have gotten so far out of date. This is particularly true for GIS databases built on digital orthophotographs and the capturing of the visible planimetric data from those photographs. For local, regional, and state government and most public utilities this resource is a central part of the database, and there must be a procedure for replacing it. There is a point at which the photography is so out of date that it becomes difficult to make decisions based on it because no one has any belief that it represents the real world anymore. So, your organization really has no choice about whether to update, just choices about how frequently and the spatial extent of updating.

Before GIS implementation most users of data gathered from aerial photography would update the entire service area on a 5- or 10-year cycle. The delivered products were sets of negatives and photographic prints. Now the deliverable products are digital orthophotographs and vector information such as road centerlines, vegetation, fences, utility poles, and so on. You could draw new information on the old photographs, but it was not very helpful. With GIS you will have processes to add updates to the spatial information but not to the air photo. So eventually you will have to reacquire photographic or satellite data to place behind your other information. If your service area is large, it is possible to rotate data acquisition around the area and spread out the total costs by reacquiring only a portion of the service area each year. If your service area is small, however, you will probably have to adopt a reacquisition policy that covers the entire area. The frequency of this reacquisition depends on how much change you are expecting in the region and what resources you have to devote to it.

As more organizations in your service area begin to use GIS and need this critical resource, it becomes possible to leverage resources by having organizations band together to reacquire the data. Utilities and municipalities cooperating on this kind of data updates are common. There are even instances where utilities have contracted for the aerial photography and data acquisition and then make the data available to other organizations in the region or state on a cost recovery basis with no prior agreements. There are significant economies of scale in aerial photography and the subsequent processing, so an organization that can afford up front to do it for its own purposes may be able to resell that information to other users. And as the satellite information gets better and more spatially detailed, more GIS users will migrate to that data source. It is even possible to adopt an update strategy of acquiring the expensive, highly detailed photographic information infrequently and updating it in selected areas with less-expensive, less-detailed satellite imagery. Being creative on how and when to update the expensive elements of your GIS database can make a huge difference in its utility to the organization.

Other GIS implementations may rely on updates from government agencies or data vendors that enhance and resell that data. GISs that are built principally to route deliveries or people need to ensure that every possible road segment with correct address ranges is in the database. Although a local government may be able to get by with 5-year updates of aerial photography, a regional delivery or public safety organization may need quarterly updates of their database. It is dangerous and potentially very expensive from a liability perspective to have a service call come to an ambulance dispatcher who is unable to find the address. Frequent and accurate updating of this kind of layer in a GIS is absolutely vital, so you need to plan for it. Usually there are regular releases of this type of data.

The final part of a maintenance plan is a schedule and process for evaluating the GIS. Although among those who were advocating the GIS there may be a feeling that it is meeting needs and everyone is completely satisfied with it, higher-level management may need more reassurances than that. Timing of an evaluation is important: too soon and people are not used to it, not rapid with their work, and still tied to the pre-GIS way of doing things. Do the evaluation too late and they may have forgotten the tedium of the tasks that the GIS has freed them from and be complacent. Six months to a year after implementation is a good time to do an assessment of the system. This user and manager survey is also a good opportunity to get information on additional applications users might need, suggestions for training, additional data, and other aspects of the system. Evaluation is always a risky activity because there is a possibility that users are quite dissatisfied with the system and how it was implemented. However, it is a necessary step and always part of the development cycle for a new or existing system.

Data Dissemination

Ownership of GIS layers and feature data sets inside the organization is determined by who needs to use the information and what kinds of roles and privileges they need to do their jobs; it is internal to the organization and driven by its needs. When ownership is examined from outside the organization, legal issues of copyright, licensing, and access arise. If you separate issues of management from those of design and implementation, legal concerns are a management issue, complex and closely linked with legal issues around all kinds of information. But in the design process the legal concerns are simpler. If some state or local statute requires that certain information be kept about geographic features, your database design needs to include space for that information. Highway and engineering departments are required to keep track of the road signage, where it is, what kind of sign it is, when it was installed, and so on. Accident liability can be a significant cost for a government. These legal concerns in the design process, however, involve the details of the table and not organizational design.

Inside the Organization

One organizational issue that does have substantial legal implications is the establishment of procedures to disseminate the GIS data, both within the organization and outside it. Within the organization, decision-making users need to have timely information that is adequate for their needs. If the organization fails to deliver that information and a poor decision is made, there can be significant consequences for the organization. Decisions on how to make the information available to all users within the organization need to come early in the design process because they have significant implications on network and database design and on hardware. There are many options for how to do this, but they all come down to suboptions or mixes of the following options:

- Continuous, real-time interaction between the user and the database, which is maintained centrally.
- Regular updates of the entire database or portions of the database delivered to the users to reside on their computers. Delivery can be through a network or by some storage medium such as a CD or portable disk drive.

Choosing to have any real-time access for users (not maintainers who, of course, will need direct access) strongly suggests that the unit responsible for maintaining the database be capable of maintaining not just the database but the hardware and software needed to make certain that this access is secure and continuous. These are not small issues, and it would almost require that your GIS database and most of its maintenance activities be located in an IS or IT department. If you select the regular update methodology of delivering data to users, the unit responsible for maintenance and distribution needs only the database and GIS skills, and you avoid the problems of maintaining servers and networks for distribution. The downside of this approach is that data are possibly of out of date by the time users get it. If you are constructing a GIS database that changes infrequently, the physical distribution option makes more sense. A mixed approach would replace the distribution by physical storage media such as a CD with automated distribution over a network. In this hybrid situation, a unit that has the database and GIS skills maintains the database, and copies of the database or portion of it are transmitted to users over a network; the procedure would delete the existing data and replace it with the newer data. This avoids problems of multiple users or modifiers trying to access and work on a single database installed on a server. In situations where users are widely separated and not on a single network, some kind of physical movement of data will be necessary.

Another internal distribution method to consider is through an intranet application. With Web programming it is possible to push data from a server to multiple desktop computers. Network administrators routinely update software across a network this way, and it can be used to move files as well. It is also possible to make the data accessible to users on a Web site; this is an increasingly common way to make geographic data available to the public, but it can be used within the organization as well. If the access to data updates is simply made available and not physically moved onto the user machines, you must rely on the users to obtain and install the data updates, and some will not. This creates the problem of some people and units working with older versions of the data. This can cause a lot of confusion, so any Web-based distribution needs to physically move and install the data on the correct machines.

What is likely to happen is that there will be multiple means of making the database available to the users and maintainers of the data. The details of exactly what data goes to what people how often is determined in the implementation phase, but a decision of how to distribute the data within the organization is an important task in the design process. It determines how the GIS will mesh with what you need to do.

Outside the Organization

It does not matter whether you are a private business, a governmental organization, or a nonprofit; once you create a GIS database, someone else will want a copy of all of it or pieces of it, and you need to be prepared for how your organization will deal with these requests. You will spend a lot of time thinking about this and doing it once you have figured out the process. This is actually part of the design process that can wait because it does not become a concern until the database is implemented. After all, you cannot distribute data you don't have. If your organization is a private business, your decisions appear to be simpler than if you are a governmental organization. If, as a private business, you develop geographic data sets that help you do your job and make profits, you do not have to provide copies of that data to anyone if you do not wish to. If you do decide to make them available, you can put whatever price on it you want, and potential users will either purchase it or not. What is increasingly common among for-profit GIS data providers is to license the use of the data rather than directly selling the information. The license agreement will specify what the licensee may and may not do with the data and what the responsibilities are of the licensor with respect to the data (e.g., delivery of updates, answering technical questions about the data). The data are actually delivered, and the license buyer has a copy of them, but the use of the data is governed by the license agreement and, theoretically, if the user violates that agreement, there are penalties, the data must be returned, and so on. The GIS data of a private firm is a product, and the firm may market it as it does any product.

In the public sector data distribution gets more complicated because public money was the funding source for the development of the data. In the United States, under the Federal Freedom of Information Act and the state equivalents,

Open Records Acts, the public must get access to this information somehow. They have a right to see the data and in many case a right to obtain copies of the data. The case law in this area is large and frequently changing, so legal advice is important before you start distributing data. After liability concerns, any organization that is distributing data, either by choice or necessity, is concerned with how to price that data and who will bear the costs of distribution. There are three types of cost structures in data distribution, whether or not it is geographic data:

- *Profit making.* Only private businesses may make money selling their data. Governmental units in the United States are not allowed to make a profit on data sales.

- *Cost recovery.* This costing model requires the provider to calculate the costs of building the data set and estimate the number of potential users over the life of the data set, and by dividing the costs by the estimated users arrive at a cost per user. This would be what you would charge the buyer. In Europe, governments are allowed to charge for data this way. In the United States it is more confusing, so you should definitely seek legal opinion before implementing a cost structure of this sort.

- *Cost of duplication.* It is clear that governments may charge for duplication costs and have been doing so with information for many years. You may include staff time, pro-rated hardware and software maintenance, supplies, and clerical time to process the request. Some governments set up fees based on the size of the computer files or the medium the data are copied to. For example, the USGS charges a set amount per compact disk of data regardless of what is on the disk.

GIS Data Distribution through the World Wide Web

Of course you can always make the decision to give your data away to anyone who asks. Increasingly, organizations that take this approach to outside distribution of their data are using the World Wide Web to make their data available. There are hundreds of Web sites, mostly state and local governmental units, that make their data available to all users this way. It still requires staff time to set up the distribution Web pages and to replace old information with updates, but it is a very efficient way to distribute the data. If you choose to make your data available either on a fee basis or at no cost, there are some concerns you need to deal with:

- *Metadata.* Metadata should always be available for any information you place on the Web. Ideally it should be incorporated directly into the data or the file that the users will download rather than having a separate button they have to click to get the it. Distributing metadata along with the GIS data is, by itself, a good thing to do, but the real reason you should do this is to save the time it might take to answer the questions people will have about the data. Good metadata should describe the data so well that there is no need to e-mail or telephone the distributor about the data.

♦ *Disclaimers.* There are always liability issues around data released by organizations. This concern should not stop you from distributing your data to people who need them so long as the users understand the data and what the appropriate uses of the data are. Table 10.4 contains some sample disclaimers.

Disclaimers such as these have their origins in the disclaimers that organizations have been putting on maps for generations. They serve as a warning to the users in sufficiently legal terms that they are receiving an as-is product with no warranty or guarantee about its accuracy and, most importantly, does not place the organization at any risk for use of the information.

♦ *Update frequency.* If you are distributing via the World Wide Web, it is not going to be possible to have real-time, at the moment, access to the data. Periodically you will make a copy of the data you want to distribute and place it on the Web site. It is a good idea to plan for the frequency of the updates and explicit state the currency of the data in the metadata. If you are going to refresh the Web site data frequently, it makes sense to construct specific applications that will automate the process. But outsiders will probably be working with information that is not as up to date as the data to which the workers in the organization have access.

♦ *Data formats.* As anyone who has worked in GIS for some time knows, there is a bewildering array of data and file structures that different software systems use for GIS data. There is no way any organization could, or should, make their Web-based data available in all the possible formats for the occasional user who might need them. Generally, Web-based data access is provided for knowledgeable GIS users who can transform the data from whatever format(s) you select into the format that meets their needs, and that should be their responsibility, not yours. Casual browsers will need map-like interfaces rather than data access. Table 10.4 lists some the common formats used for posting GIS data on the Web, with some discussion of the issues they raise.

Generally, it is easier to use exchange formats that take an entire layer of data and deconstruct it into a single file. You also need to consider the abilities of other GISs to be able to read or import the data. This is why the Ungen format of ESRI, which is structurally the simplest format in the list, is not recommended for Web posting; users will need the more expensive core GIS software from ESRI to be able to read and use the data. You want to provide data in a format that will require the least processing on the user's part to be able to view and use them. The following are factors to consider regarding Web deployment of your GIS:

Table 10.4 Samples of Disclaimers

California State Lands Commission:

March 12, 1995 This data has not been approved by the California State Lands Commission and does not constitute an official map or dataset from such commission, nor does this data establish the boundary lines of any state owned lands depicted there- on. This data is in preliminary form only and is subject to change without notice. The California State Lands Commission makes no representation or warranties with respect to the contents hereof and specifically disclaims any implied warranties of merchantability or fitness for any particular purpose. The name of the California State Lands Commission may not be used in advertising or publicity pertaining to distribution of the dataset without specific, written prior permission.

gis.slc.ca.gov/slcgis/data/disclaimer.asp

Federal Communications Software, on its GIS page:

The Commission makes no warranty whatsoever with respect to the software. In no event shall the Commission, or any of its officers, employees or agents, be liable for any damages whatsoever (including but not limited to, loss of business profits, business interruption, loss of business information, or any other loss) arising out of or relating to the existence, furnishing, functioning or use of the software.

uls-gis.fcc.gov/

Institute of Water Research, Michigan State University:

The Institute of Water Research is NOT responsible for the accuracy of all data available in this web site. All data were converted from original sources to Arc/Info(c) "coverage" format and projected to the same coordinate system. All data are provided "as is" and without warranty of any kind. The Institute of Water Research will NOT distribute any data available in this web site. Please do NOT request any data. However, any comments and suggestions are welcome.

gis.iwr.msu.edu/gisdisclaimer.html

Nashville Planning Department:

The Nashville Planning Department of The Metropolitan Government of Nashville and Davidson County presents the information on this web site as a service to the public. We have tried to ensure that the information contained in this electronic document is accurate. The Planning Department makes no warranty or guarantee concerning the accuracy or reliability of the content at this site or at other sites to which we link. Assessing accuracy and reliability of information is the responsibility of the user. The Planning Department or The Metropolitan Government of Nashville and Davidson County shall not be liable for errors contained herein or for any damages in connection with the use of the information contained herein.

nashville.org/mpc/disclaimer.html

Camarillo, Calif.:

GIS maps created by the City of Camarillo Geographic Information System, are designed for the convenience of the City and related public agencies. The City does not warrant the accuracy of these maps, and no decision involving a risk of economic loss or physical injury should be made in reliance thereon.

2.ci.camarillo.ca.us/govt/legal.html

- *Interface design.* If you have ever cruised the World Wide Web looking for geographic information for a project, you will appreciate the different interfaces people have developed for data access. Some clearly work better than others, and some seemed designed to hide the data rather than make it easy to find. Organizations are usually quite sensitive about how they present themselves to the world through the Web, and a well-designed interface is important. For the liability-conscious it is not difficult to place disclaimers in the interface so that the user has to click to get by the disclaimer, thereby registering and documenting that they have seen the disclaimer. This is just like the many license and use agreements that are on the Web for software and other services; most people don't read them and just click by. They are the Web equivalent of the fine-print product information on packages and are a good idea.

- *Provide an image of the data or not.* It is common, but not universal, for providers to supply a graphic image of what the data will look like once they are imported into a GIS or mapping program. This is a nice service, but it adds another layer to the interface. For example, if you choose to make metadata available and two different formats for the data and a graphic image of the mapped data, you will need at least four separate files for each layer. If you update frequently, this can lead to significant work.

- *Compression technology.* Fortunately, file compression has become so standardized and the software much better that there are not many decisions to make. Because most GIS layers tend to get very large, it is almost essential to compress the files before placing them on the Web. Compression also allows you to assemble multiple files, if your format requires it, and the image and metadata into a single file so that when the user clicks on the download button, the entire package will be transmitted. With the widespread diffusion of the WinZip compression utility, most PC users are using that compression program to zip, or compress, files and folders. Unix users have a slightly different set of compression utilities, but the WinZip decompression program recognizes and can deal with that type of compression. Compression is necessary to move large amounts of data around the Web, but frequently a user will be using a particular Internet provider that has restrictions of the size of files it will allow you to receive.

- *Alternate means of distribution.* Some people who will need your data may not have Internet access or their access is not adequate to deal with the volume of data they may need. For those people you will need an alternate means of distribution. Many public organizations provide the data at no cost on the Web but charge distribution costs for data that must be prepared on CD, disk, or tape. This is also the process adopted by several commercial distributors of data; people who wish to wait and do the work themselves can do it free, but if you want the company to copy and send you the data, there will be a charge. Compact disk is becoming the standard format for that kind of distribution because the technology to copy disks has become so inexpensive.

Table 10.5 Data Distribution Formats

Data Formats	File Structure	Import Concerns	Results of Search on Google (5/8/02) (in 000s)	Search Term
Autocad DXF	Single file	Topology must be reconstructed after import; all GISs will import; many free viewers available.	368	.dxf
ESRI shape files (.shp)	Multiple files, at least 3	Many GISs will import.	240	.shp
ESRI export (.E00)	Single file	Must have import software; distributed with all ESRI software products, including freeware.	107	.e00
ESRI coverage	Multiple files and folders; not recommended	Need ESRI software to read; complex file structure; use E00.	No search	
ESRI Ungen (ASCII text, ungenerated coverage)	Single file	Need core ESRI software to read; topology must be reconstructed after import.	40	Ungen
Mapinfo Interchange Format (MIF)	Single file	Most, not all, GISs will import.	7	.mif and mapinfo
Mapinfo tables	Multiple files	Not recommended, even for exchange with Mapinfo installations; use MIF.	No search	
Spatial Data Transfer Standard (SDTS)	Multiple files; complex naming system	All GISs can read but not most inexpensive viewer systems.	110	SDTS
Digital line graph (USGS, DLG)	Single file	Preserves topology; all GISs can import and export; most inexpensive viewer versions cannot.	238	DLG
Topologically Integrated and Geographically Encoded Reference System (U.S. Bureau of the Census, TIGER)	Multiple files; complex naming system	Many free viewers available.	51	Tiger and GIS

However your organization chooses to deal with these issues of Web deployment of your GIS data, the up-front setup time will quickly pay off in decreased time for staff to deal with data distribution requests. Organizations that do not wish to be as free and available with their data, of course, will not adopt Web technology or will control access to the Web distribution, which is an in-between way to control access. With this hybrid approach you can license or sell the rights to access the data pages and protect that access with a password that will let you into the Web interface.

How They Did It – GISs Data Distribution through the World Wide Web: Montana Natural Resource Information System

Organizations that want to share their geographic data widely often turn to the Web as the distribution mechanism. The Natural Resource Information System (NRIS) is part of the Montana State Library; because libraries have a data collection and distribution function, this was a natural extension of their mission. "We created our Web site because out director at that time, Allan Cox, was very interested in the World Wide Web, and he and our GIS manager, Fred Gifford, realized it would be a tremendous tool for accomplishing our mission of being a central source of natural resource information for Montana." (Gerry Daumiller, e-mail, 5/24/2002). Their IT department was willing to run a T1 network line into their building to facilitate the distribution system. In July 1994, the Web site went live, being run from a Sun SPARCstation 2 on Gerry Daumiller's desktop in Montana. The design innovation of this distribution site, which has been copied since then, was to present each layer as a row across the page and have a graphic image of the data, and buttons to download the graphic image, the metadata, and two different data formats for ArcInfo (ESRI) data:

⬚ GIF (48K) ▨ (12K) ▨ (768K) ▨ (1040K)

The second data format, shape files (SHP), was added in 1998. They included the size of the files so that you would have some idea of download time. The files were all compressed in zip format. They received a grant from the Federal Geographic Data Committee to make their metadata FGDC compliant and became a node in the National Geographic Data Clearinghouse sometime in 1995.

All their GIS data were put on the site in May of 1996, by which time they felt the World Wide Web was fast enough to handle large data downloads. The site had ups and downs during the period they monitored its use through the middle of 1999. In late 1998 their old WWW domain name, nris.mt.gov, was dropped from the Domain Name Service and Yahoo dropped them from their Web site categories. The number of shape and export files (data) downloaded dropped from over 3,000 per month to around 1,200 per month by the middle of 1999. They have found that by far most of the downloads from the site have been for the graphic images of the maps rather than the geographic data themselves. Surprisingly, there have been more downloads of the metadata files than the data themselves, but these requests were dwarfed by the number of people who just wanted a static map.

This distribution Web site demonstrates some of the problems organizations encounter when they try distributing over the Web. First, access depends pretty much on how easily search engines on the Web can find you. When Yahoo dropped the site from its map page, the number of requests dropped dramatically. Second, it may not be the data themselves that people are interested in. In this Web site the graphic map files are much more popular than the data. It appears that most casual users have needs for maps to drop into documents but do not want to or are unable to manipulate a GIS to visualize the data. A common complaint among people who manage these Web sites is the number of e-mails and telephone calls they have to answer to explain why a map does not just pop up on the screen when they click on the data button. Third, these distribution systems seem to have a life cycle of ecstatic initiation, when the organization and managers monitor the site and watch its popularity grow to maturity and when they stop updating the site and monitoring its usage.

Summary

An enterprise GIS is a complex and dynamic thing. Like an automobile or a house, favorite metaphors for a GIS, it will need careful design and construction, regular maintenance, and occasional significant overhauls and improvements. Just as a 20-year-old kitchen begins to look a little worn and not work as well as it used to, a 4-year-old GIS is beginning to show signs of age. Users will come and go with special needs and skills, and there will be constant new sources of data. As software and computer processors improve, GIS may even move away from its traditional flat map interface, and viewers may interact with something that looks like an oblique aerial view of the area of interest that they can move around and query. Staying at the cutting edge or just behind it, which is a safer location, of these changes in technology and data requires a lot of hard work, but the rewards are significant. You can do routine work more rapidly, so the cost-benefit ratios in GIS usually work out reasonably well. If your GIS is well designed, flexible, and inviting, the most significant changes that it will bring to your organization are the new questions people can now ask and the new ways of looking at the organization and the information that drives it. A GIS is an empowering and integrating technology, and almost all organizations that have developed enterprise GISs are glad they did it and appreciate the amount of work it required. Always remember that the database is at the core of your GIS. However cleverly you have designed your applications and interfaces, without useful data behind them, they give you no assistance in making decisions. Keep that core of your GIS in good shape and you can always work out the inevitable and never-ending technological and organizational concerns that will get in your way.

Like the organizations whose missions they support, all enterprise GISs are different; off-the-shelf solutions always require some modifications and improvements. But there are consistencies in the process of design, implementation, and management that the organization must address if their GIS is to do what it is supposed to do. We hope this book has assisted you in identifying issues and will help you work through them as you design and implement your enterprise GIS.

ADDITIONAL READING

Mark, D. M., and A. U. Frank. 1992. User Interfaces for Geographic Information Systems: Report on the Specialist Meeting—Report 92-3. National Center for Geographic Information and Analysis: Buffalo, NY.
ncgia.ucsb.edu/Publications/Tech_Reports/92/92-3.pdf.